The Big Idea

What Shape Is Space?

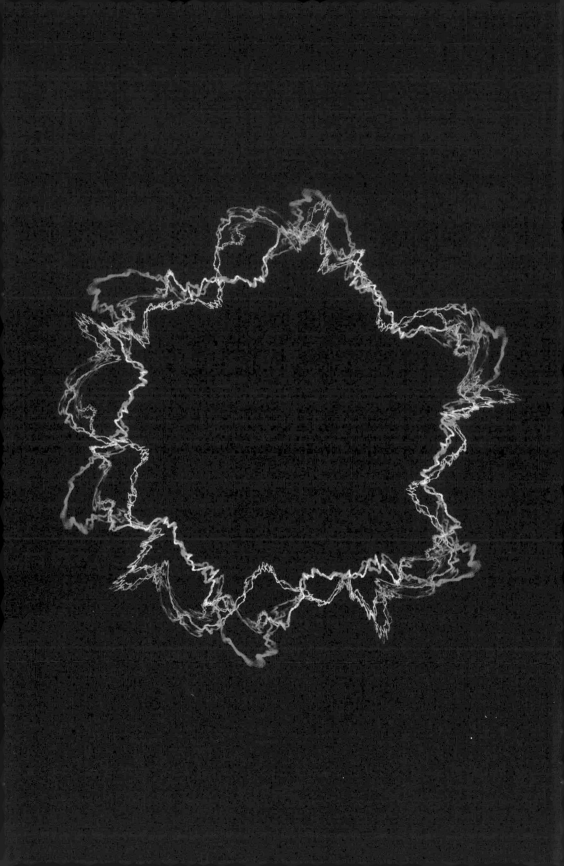

The Big Idea

Giles Sparrow

What Shape Is Space?

A primer for the 21st century

Over 200 illustrations

Thames & Hudson

General Editor:
Matthew Taylor

Contents

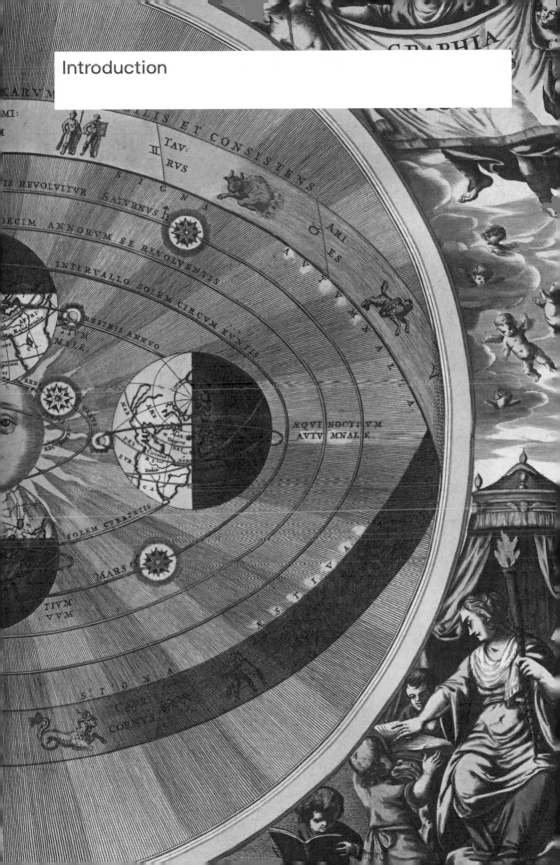

A

This book sets out to answer a deceptively simple question – what shape is space?

It is a question that matters because measuring and interpreting the shape of space has huge implications for cosmology – the science of the past, present and future of the Universe itself. This field examines evidence for how and when the cosmos began, how far it might extend today and what its ultimate fate might be, many billions of years from now.

Universe For our purposes, a 'universe' (without an upper case 'u') is simply a continuous region of space across which we could plausibly travel. 'Universe' with an upper case 'U' will be used when referring specifically to the Universe in which we reside.

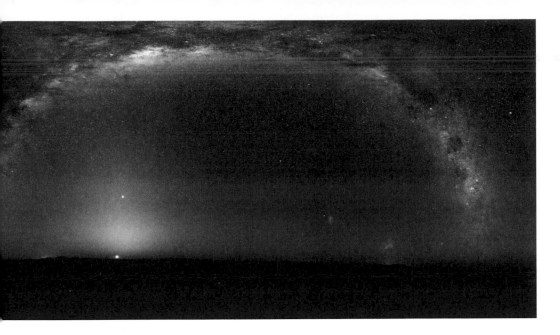

A   A sky full of stars hangs above the enormous buildings of the European Southern Observatory's Very Large Telescope on Cerro Paranal, Chile. These monster telescopes sweep up light rays from the furthest corners of the cosmos at the end of a journey that may have taken many millions, even billions, of years, thereby providing clues that help us to understand the structure and shape of the Universe itself.

But before we look at the various lines of evidence indicating that space has a particular shape, and the different theories that can be used to understand it, we first need to understand exactly what cosmologists mean when they talk about space, and how it can have a shape at all.

When we think about space today, the first picture that comes into our heads is probably *outer* space: the vast empty void that begins somewhere above us at the edge of the atmosphere, and stretches across enormous emptiness, separating the planets and stars and filled with nothing in particular.

Space, as we imagine it, is the cold, empty, inky blackness between the stars.

But for physicists and cosmologists, space has a rather different definition, one that we encounter just as much on Earth as in the wider Universe. In this conception, space is the geometric structure that provides a frame of reference for making measurements: objects occupy volumes in space, and move through it from one position to another.

Intuitively, we perceive space as having three dimensions. We measure objects in terms of their height, length and depth, and similarly understand their movement: up and down, side to side, and back and forth. We appreciate that these are different measurements, but also that they are interchangeable depending on our point of view; if you and I are looking at the same object in different orientations, we may disagree on which direction we measure as its height, its length and its depth.

One thing we will agree on, however, is that the three directions are perpendicular to each other, that is, set at right angles of 90 degrees. What's more, if we can agree on a single fixed point of view called an origin, on a common orientation and on a shared unit of measurement, we can define any point in space in terms of only three numbers, indicating those three measurements. These simple principles form the basis of the Euclidean geometry we all wrestled with in school.

Dimensions are simply a scientific term for independent directions of measurement. Hence, we can say that the space described by Euclidean geometry is three-dimensional. We might, if we had enough string, map it

A

Euclid This Greek mathematician wrote the earliest surviving geometry textbook, the *Elements*, around 300 BC. In it, he set out various axioms of geometry on flat planes, which in fact apply equally well in three-dimensional space.

A   This well-known engraving from an 1888 book by French astronomer Camille Flammarion captures the idea of astronomy as a succession of revelations of new and more complex Universes.
B   Euclid's *Elements* shows how various apparently reliable aspects of reality, such as geometry, are described by the language of mathematics. Today, we know that Euclidean geometry is just one special form with inherent assumptions and limitations.

TO inscribe a circle in a given square.

Make ——— = ·········· ,
and ——— = ·········· ,
draw ········· ‖ ———·········· ,
and ·········· ‖ ·········· .
(B. 1. pr. 31.)

∴ ▪ is a parallelogram ;
and since ——— = ·········· (hyp.)

∴ ——— = ··········

∴ ▪ is equilateral (B. 1. pr. 34.)

In like manner, it can be shown that

▪ = ▪ are equilateral parallelograms ;

∴ ·········· = ——— = ·········· .

and therefore if a circle be described from the concourse of these lines with any one of them as radius, it will be inscribed in the given square. (B. 3. pr. 16.)

Q. E. D.

B

IN a given circle to inscribe an equilateral and equiangular pentagon.

Construct an isosceles triangle, in which each of the angles at the base shall be double of the angle at the vertex, and inscribe in the given circle a triangle equiangular to it ; (B. 4. pr. 2.)

Bisect ◢ and ◣ (B. 1. pr. 9.)

draw ———. ———. ——— and ········· .

Because each of the angles

◢ , ◢ , ◢ , ◣ and ◣ are equal,
the arcs upon which they stand are equal, (B. 3. pr. 26.)
and ∴ ———, ———, ———, ——— and ·········· which subtend these arcs are equal (B. 3. pr. 29.)
and ∴ the pentagon is equilateral, it is also equiangular, as each of its angles stand upon equal arcs. (B. 3. pr. 27).

Q. E. D.

r

out as an endless grid-like pattern of cubes, with each corner marked by the intersection of three lines of string at perpendicular angles. According to Euclid's theory, this grid should remain uniform and unaltered regardless of what actual objects are placed within it, whether those objects are ping-pong balls or planets.

But in fact, Euclid only got things partially right – the presence of truly massive objects such as stars and planets *does* make a difference, and the reason that they do is because thus far we have been missing an important element of the story.

To understand what is really going on, we must first investigate the other meaning of space: as the void between stars and planets.

A

For most of human history, even the finest minds saw no need to imagine gaps between the stars. Ancient theories based on the limitations of naked-eye observation (the only option available at the time) reached the understandable conclusion that Earth is the static centre of the Universe, with the Sun, Moon and planets circling it and the stars forming an outer shell around the heavens that rotated once each day.

This explained why solar system objects move against the background stars at different rates, and why all the celestial objects rise and set once each day. Following the maxim that 'nature abhors a vacuum', meanwhile, it was generally agreed that planetary orbits and the gaps between them were filled with a mysterious substance called the aether, which allowed light and forces of motion to be carried between celestial objects.

Although some early Greek philosophers made surprisingly good estimates of the size of the Earth and the distance of the Moon and Sun, they still assumed the gap between these objects was filled with something. Only one group, the Stoic philosophers, who flourished from the 3rd century BC, imagined a Universe in which there was such a thing as a void, although their theory was that the void lay *beyond* the visible cosmos (including the stars), rather than permeating it. Furthermore, the Stoic view was rapidly displaced by another model that seemed far more successful at explaining the movements of the planets in the sky.

This theory, developed by Ptolemy of Alexandria in the 2nd century AD, envisaged Earth at the centre of the Universe, with other planets circling around it on fixed spheres composed of the mysterious aether. According to Ptolemy, the planets themselves actually sat on smaller circular tracks called epicycles, whose centres were pivoted on the principal spheres. This explained the mysterious phenomenon of 'retrograde motion', in which Mars, Jupiter and Saturn reverse their paths across the sky for a few weeks or months.

Ptolemy's model envisaged the system of Earth, Moon, Sun and five known planets encased in an outer sphere carrying the fixed stars, with no allowance made for empty space between them. Its appeal lay in the fact that it seemed to offer a reasonably accurate description of actual planetary motions, while at the same time preserving a deeply embedded philosophical idea (going back at least as far as Plato) that heavenly perfection required purely circular motion.

Ptolemy of Alexandria
(c. AD 100–170)
This Greek-Egyptian mathematician wrote an influential astronomical treatise known from its Arabic translation as the *Almagest*. He established ideas about astronomy that went largely unchallenged until the Renaissance.

B

A

The geocentric (Earth-centred) Ptolemaic system was to dominate astronomy for more than a millennium, outlasting the Roman Empire and becoming an established part of Church teaching, before new discoveries began to seriously undermine it.

It was in the early 16th century, amid the intellectual ferment of the Reformation, that Nicolaus Copernicus first dared to circulate his heretical ideas about a heliocentric or Sun-centred system.

Copernicus replaced Earth with the Sun at the centre of the solar system, with the planets Mercury, Venus, Earth, Mars, Jupiter and Saturn in orbit around it, and the Moon on its own track around Earth. This simplified many of the puzzles of planetary motion (retrograde motion, for example, became an effect produced when the faster-moving Earth 'overtakes' a more distant and slower-moving planet on its annual track around the Sun), but it did not resolve them all perfectly, and by the time of his death in 1543, Copernicus had 'fudged' his model by incorporating his own system of epicycles, similar to those used by Ptolemy to maintain the concept of 'ideal' circular motion. Nevertheless, *De Revolutionibus Orbium Coelestium* (*On the Revolutions of the Heavenly Spheres*), published on his deathbed, sparked a revolution in astronomy.

B

At first, the Copernican theory met opposition for its attempt to uproot deeply held assumptions about the nature of the heavens. In the 1570s, however, two events occurred that challenged these assumptions more directly. First, in 1572, a bright new star appeared for several months in the constellation of Cassiopeia. Now known to be an exploding star or supernova, this enormous eruption showed that the fixed stars were not in a state of fixed perfection as had once been thought.

A   A plate from Andreas Cellarius's work *Harmonia Macrocosmica* (1660) shows the traditional cosmological model of an Earth-centred Universe, surrounded by orbiting planets and an outer sphere of stars.

B   A second plate from *Harmonia Macrocosmica* shows the Copernican model of a cosmos, with the Sun at its centre and Earth as one of several planets. Note the single Moon circling Earth and the four accompanying Jupiter.

Nicolaus Copernicus (1473–1543) This Polish Catholic priest and astronomer made observations of planetary motions that led him to formulate a Sun-centred theory of the Universe. He first circulated his ideas in a small private treatise of 1514, but did not publish in full until 1543.

Constellation Traditionally a pattern made up by human stargazers out of bright stars, often with a story or legend attached. Today's constellations, however, are regions of the sky surrounding these traditional patterns; there are 88 of them and they interlock to cover the entire sky.

A

Then, in 1577, a bright comet appeared in the heavens for several months. Observations from different parts of Europe showed that its direction did not significantly change depending on from where it was seen, and so it must be immensely distant rather than an atmospheric phenomenon. What was more, the comet's path through the sky clearly indicated that it must have been passing straight through the supposed planetary spheres.

The final breakthrough came in 1608, when Johannes Kepler realized that the motions of the planets through the heavens were better explained if one abandoned the idea of 'perfect' circular motion in favour of orbits that are ellipses (ovals), with the Sun at one of two 'focus' points. Kepler's new model of the solar system caught on rapidly, and was soon bolstered by observations made with the newly invented telescope by astronomers such as Galileo Galilei.

Kepler's elliptical orbits got rid of tracks and spheres, and because his theory provided a direct link between the time a planet takes to complete an orbit and the size of its orbit, it also revealed that the planets were adrift in a vast expanse of empty space at distances of tens or even hundreds of millions of kilometres. And as we shall see, the realization that Earth orbited the Sun carried an even deeper implication about the distance of the stars themselves.

The scale of the Universe was hugely expanded in one fell swoop, and for the first time in history, astronomers were forced to get to grips with the idea that Earth is surrounded by a vast gulf of space extending in all directions. Yet it was still possible to impose some sort of shape on this void, and key to this was mapping and understanding the distribution of objects within it.

Johannes Kepler (1571–1630) This German mathematician and astrologer served as assistant to the Danish astronomer Tycho Brahe before becoming imperial mathematician to the Holy Roman Emperor Rudolf II. Access to Tycho's observations of Mars convinced him to formulate his laws of planetary motion in the *Astronomia Nova* of 1609.

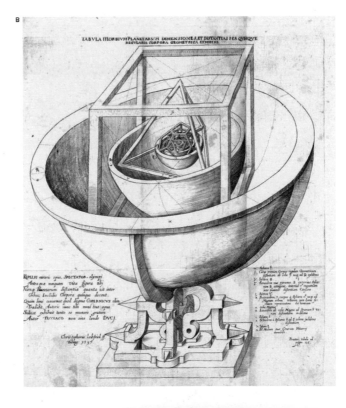

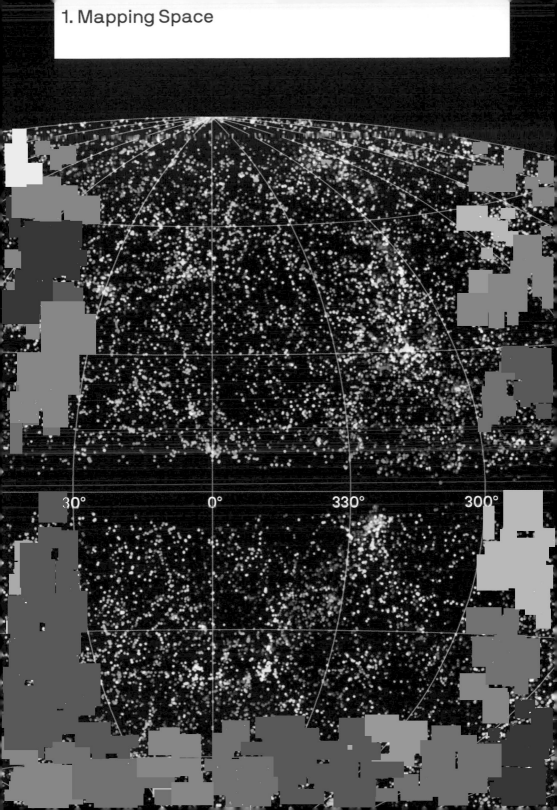

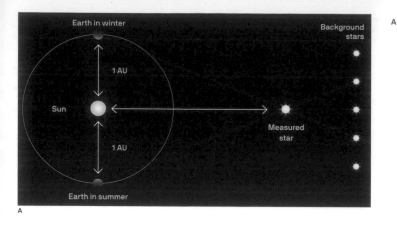

A Stellar parallax arises as Earth moves from one side to another of its orbit. When seen from either side of this 300-million-kilometre (186.5 million mi) 'baseline', a nearby star will shift its direction in the sky, and its apparent position compared to more distant background stars. Yet such is the scale of the Universe that the shift is tiny for even the nearest stars.

The story of our modern quest to understand the shape of space really begins in the Copernican revolution that unshackled celestial objects from the limits of crystalline spheres and allowed them to roam free in a vast gulf of space. Uprooting Earth from its hallowed place at the centre of the Universe and (for a time) elevating the Sun to cosmic pre-eminence had important and immediate implications for the distance of not only the other planets, but also of the stars themselves.

> For centuries, a key argument against the heliocentric system had always been that if Earth was moving in space, the direction of the stars would surely appear to change as our point of view shifted from one side of Earth's orbit to the other.

This effect, known as parallax, is very familiar from our everyday lives: hold up a finger at arm's length and wink each eye in turn, and its position will appear to shift back and forth when compared to more distant background objects. Astronomers had tried and failed to detect such a parallax shift in the stars, and the lack of any evidence had weighed heavily against the idea of a moving Earth.

But as telescopic discoveries and the power of Kepler's system brought overwhelming new evidence for Earth's motion, the same argument flipped into reverse. If the stars did not appear to change their apparent directions even as the Earth moved some 300 million kilometres (186.5 million mi) from one side of its orbit to the other, then they must be *unimaginably* distant.

The quest to measure parallax occupied many astronomers through the 18th and 19th centuries. In theory, the increasing power of telescopes made the task easier, but in practice there were still major challenges to overcome in terms of 'astrometry' (the measurement of stellar positions). With the stars wheeling around the sky once every 24 hours and the Sun rising and setting, we cannot simply keep a telescope trained on the same point in space and observe if a star moves gently back and forth; instead we have to measure stellar positions on a fixed grid of 'celestial coordinates' (akin to latitude and longitude on Earth).

One early discovery was that the stars are not as fixed as they appear to be; many of them move slowly across the sky from year to year and decade to decade. This so-called 'proper motion' is caused by a combination of the star's motion and that of our own solar system, and it varies considerably from star to star. Astronomers soon realized they could use this as a likely indication of which stars are closest to Earth, and focus on those for their attempts at measuring parallax.

B   Proxima Centauri, the closest star to Earth, lies only 4.25 light years away – a distance at which it shows a parallax shift of 0.77 seconds of arc (about 1/2000th the width of the Full Moon) and a proper motion that sees its position shift by a Full Moon's diameter every 450 years. Despite its proximity, it was not discovered until 1915.

C   This detailed view plots Proxima Centauri's motion against background stars through the mid-2010s (caused by a combination of both Proxima's motion and that of Earth). The 'loops' in the path are due to parallax.

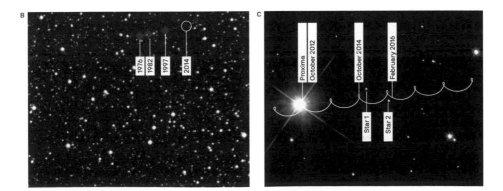

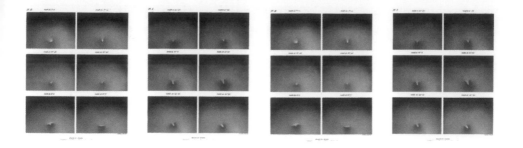

A

Nevertheless, it took many years of unsuccessful attempts before Friedrich Bessel successfully measured the parallax of a star called 61 Cygni in 1838. The yearly shift in the star's apparent direction proved to be a tiny 1/11,500th of a degree, but this was enough to calculate that it lies some 100 million *million* kilometres (61 trillion mi) from Earth.

Space had expanded once again, to a scale where everyday distance measurements became nonsensically huge.

Friedrich Bessel (1784–1846) This German astronomer accurately charted the positions of more than 50,000 stars. Taking into account the bending of light by the atmosphere and the slight changes in the direction of stars caused by Earth's motion around its orbit, he finally measured parallax successfully in 1838.

Degree A unit of angular measurement. There are 360 degrees in a circle and 90 degrees in a right angle. A degree can be split into 60 'minutes of arc', each of which can be subdivided into 60 'seconds of arc'.

Light year The distance light travels through a vacuum in one year, equivalent to 9.5 million million kilometres (5.88 million million mi).

Luminosity A measure of the total energy emitted by a star or other celestial object compared to the Sun. Luminosity differences can be used crudely as an indication of relative brightness in visible light, although hot or cool stars can emit most of their energy in the invisible ultraviolet or infrared.

Consequently, astronomers rapidly adopted the idea of measuring a star's distance from us in terms of the amount of time its light, the fastest thing in the Universe, takes to reach us travelling at an astounding 299,792 kilometres (186,282 mi) per second; on this scale, 61 Cygni is 10.3 light years from Earth.

An alternative way of measuring cosmic distances is in terms of parallax seconds or parsecs. One parsec is the distance at which a star would have to lie in order to display a parallax of 1 second of arc (1/3,600th of a degree), equivalent to 3.26 light years. Modern astronomers prefer to talk of parsecs (and indeed kilo- and mega-parsecs) rather than light years because of a preference for directly measured over 'derived' units, but for the rest of this book we will use the more familiar terminology.

One immediate discovery arising from Bessel's parallax measurement was confirmation that not all stars are the same. 61 Cygni is relatively faint (on the edge of naked-eye visibility, in fact), and once its distance was known, it was clear that it was far less intrinsically bright (less luminous) than the Sun. In fact, 61 Cygni is a 'double star', a pair of stars close together in the sky, with one slightly fainter than the other. The discovery that they both displayed the same parallax and therefore really were at the same distance also proved that these two stars must be physically different from one another because otherwise they would appear to have exactly the same brightness.

This and other discoveries around the same time opened the way for the science of astrophysics and our modern understanding of how stars work.

0°01'

A   Friedrich Bessel made these sketches of Halley's Comet during its passage close to the Sun in 1835. He had previously made new calculations of its orbit to more precisely predict its return to the inner solar system.
B   Despite their closeness in the sky, the binary stars of the 61 Cygni system are actually separated by a distance only slightly smaller than Neptune's orbit around the Sun, and take 659 years to orbit each other.

From the point of view of the growing scale of space, the most interesting consequence of this work was the realization that laborious parallax measurements are not always needed to estimate the true characteristics of a star; there are clues hidden in a star's light, such as the colours of its energy output and the signatures of elements in its atmosphere, that can reveal its true physical properties. These are often enough to give a ballpark estimate of the star's true luminosity, allowing us to estimate its distance based on its brightness as seen from Earth without direct measurement.

Hence, astronomers began to discover that many apparently faint stars, often only visible through telescopes, were in fact physically very luminous and must therefore be vastly more distant from the Sun, at distances of thousands or even tens of thousands of light years from the Sun. Finally, they were getting to grips with the true scale of our galaxy, the Milky Way.

A   Galileo published his telescopic observations of the heavens (such as the Pleiades star cluster) in 1610 in the pamphlet *Sidereus Nuncius* (Starry Messenger).
B   Galileo's sketches of the Moon revealed craters and mountains.
C   The pale band of the Milky Way resolved into countless stars, clustered together in clouds that continued far beyond the limited abilities of Galileo's telescope.

c

The Milky Way is one of the most obvious features of Earth's night sky (at least from dark locations) – a band of pale light stretching from horizon to horizon and forming the background to some of the brightest constellations. Known and wondered at since ancient times, it was an obvious target for the first telescope of Galileo Galilei in 1609. The Italian genius found that it simply consisted of many millions of previously unseen stars, stretching away to the limits of visibility.

Looking at the distribution of stars in the sky and their concentration in the Milky Way soon led astronomers to the conclusion that our solar system is embedded in a flattened plane of stars, much wider than it is deep, and with its densest regions in the direction of the constellation Sagittarius. When we look across the plane of this galaxy, the stars 'pile up' behind each other in similar directions, merging to form the clouds of the Milky Way, but when we look out of the plane, we see only a few relatively nearby stars and the darkness of apparently empty space beyond.

A

A   In the mid-19th century, astronomers resolved the spiral structure of many distant nebulae for the first time.

B   A boom in popular astronomy books such as *Astronomy of Today* (1909) coincided with a period of huge scientific discovery.

C   While early photographs showed that some nebulae contained small numbers of stars embedded in gas and dust, others appeared to be dominated by truly vast numbers of unresolvable stars.

Early attempts at mapping the scale of the Milky Way suggested that it was tens of thousands of light years across (it was only in the 20th century that its true form, a vast spiral 100,000 light years across with a huge bulge of stars at its centre, became apparent). Today, we know that Earth lies some 26,000 light years from the very heart of the Milky Way, and that everything in our galaxy is ultimately in orbit around a vast black hole with the weight of 4 million Suns.

By the early 20th century, we had come a long way in our understanding of the scale of space, vastly diminishing our own planet's perceived status in the process. But there was one further shock to come. Most astronomers of the time believed that the Milky Way was effectively the entirety of the Universe – a city of stars floating alone in a vast black void. There were few cosmological theories to explain how such a system came to be, or to describe what lay beyond the bounds of the Milky Way. Did empty space continue forever or was the entire Universe a bubble of space not much larger than our galaxy?

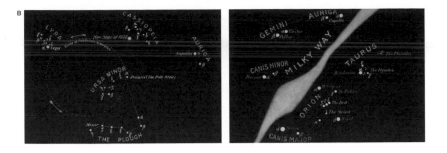

A few influential voices, however, argued that this 'island Universe' theory was mistaken. They pointed to mysterious 'spiral nebulae', faint, fuzzy blobs of light, sometimes with a pinwheel-like structure, that were visible through powerful telescopes. The conventional view was that these spirals might be new stars and solar systems at various stages of formation, but the doubters pointed out that if this were the case, we might expect them to have a similar distribution to existing stars, within the plane of the Milky Way. Instead, the spiral nebulae were largely found in those relatively sparse regions of the sky where we look past a scattering of stars into the deep void of space. Perhaps, then, these were galaxies like our own, many millions of light years distant?

This so-called 'Great Debate' simmered in astronomy for a couple of decades before a breakthrough finally came in 1925. Using the giant new telescope recently built at Mount Wilson, California, Edwin Hubble took photographs that resolved some of the larger and brighter spirals into individual stars. What was more, he identified some stars that pulsated regularly in their apparent brightness.

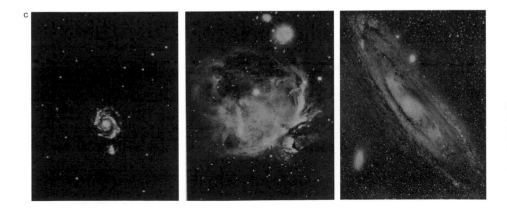

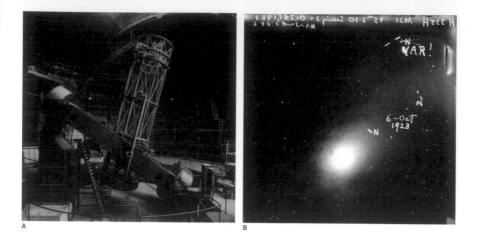

A few years previously, US astronomer Henrietta Swan Leavitt (1868–1921) had successfully identified one particular class of such pulsating stars, known as Cepheids, in which the period of brightness variation is directly linked to the star's intrinsic luminosity. After identifying Cepheids in his spiral nebulae, Hubble was able to calculate their true brightness and therefore their distance from Earth, confirming for the first time that the spirals and related nebulae are indeed galaxies in their own right.

What was more, since the vast majority of galaxies were simply too faint and distant to distinguish any separate stars within them, it was clear that the Universe must extend for hundreds of millions, even billions, of light years in every direction.

From the Copernican revolution to Hubble's breakthrough, the scale of the Universe, and space itself, had expanded from a few thousand kilometres to many billions of light years. The same journey of discovery had seen our own planet diminished in status from the centre of everything to a tiny speck of rock wheeling around an insignificant star, one of two hundred billion or more in a Milky Way that is itself one of perhaps a million million galaxies. We will see later that even this view of the Universe may be just a glimpse of the whole, but for now it is enough to understand that space is a vast three-dimensional extent that goes as far as we can see, and beyond.

What, then, does it mean to ask if space has a shape?

This is the question that is addressed throughout this book, and it arises because there is something crucial missing from our previous description: space is not such a neat three-dimensional web after all. Instead, it is malleable what appear to be straight perpendicular lines can pinch together or pull apart on different scales, depending on a number of factors, the most important of which is the presence of matter in large amounts. This discovery lies at the heart of another great breakthrough of the early 20th century.

Einstein's theory of relativity emerged in the late 19th century from a debate in physics surrounding the apparently fixed speed of light. The fact that light travels at a truly amazing speed had been known for centuries, and hence it was little wonder that expected differences in speed were hard to detect. A few metres or even kilometres per second change due to the relative motions of a light source and its observer is easily missed when the speed of light itself is so many orders of magnitude larger. But as physicists developed increasingly accurate tests of the speed of light, they reached a point where they felt it should be possible to detect differences caused by relative motion of source and detector, just as they would for any other sort of wave.

A Working at California's Mount Wilson Observatory, Edwin Hubble used the largest telescopes in the world to identify individual stars in remote galaxies.
B One of Hubble's annotated photographic plates identifies the positions of Cepheid variables in the Andromeda Galaxy, our nearest large galactic neighbour at a distance of about 2.5 million light years.
C Pages from Albert Einstein's 'Zurich Notebook' chart his early development of the basic ideas of general relativity.

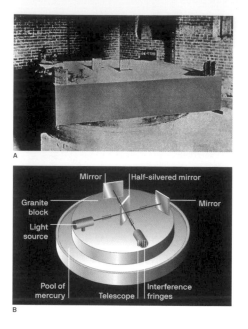

A Michelson and Morley built their experiment in the basement of what is now Case Western Reserve University in Cleveland, Ohio. The super-sensitive device sat on top of a concrete slab floating in a bath of mercury, which isolated it from vibrations and other random movements.

B The experiment is based on the principle of interferometry. A narrow beam of light is split in two by a half-silvered mirror that sends each half of the beam onto a perpendicular path. The beams bounce back and forth between mirrors before recombining at an eyepiece, creating an interference pattern as their light waves reinforce in some places and cancel out in others. If the speed of light changes along the different paths (as would be expected if the light was carried by an aether through which Earth is moving), then the interference patterns should change over time.

But no such differences emerged. The speed of light remained resolutely fixed at 299,792 kilometres (186,282 mi) per second regardless of whether a light source was moving towards or away from the detector. In 1887, US scientists Albert Michelson (1852–1931) and Edward Morley (1838–1923) devised the most intricate experiment yet to detect a change in the speed of light, and when that also failed, physics was plunged into crisis.

In 1905, failed academic Albert Einstein led the way out of the mire with a series of revolutionary scientific papers that proposed a daring solution, accepting the evidence for a fixed speed of light and rewriting all the rest of physics accordingly. The resulting theory of 'special relativity' shows that everyday phenomena remain mostly the same, and strange effects only emerge when objects and observers are moving relative to

Albert Einstein (1879–1955)
The archetypal genius physicist, Einstein was working as a patent office clerk when he wrote a series of scientific papers that revolutionized physics during the *annus mirabilis* of 1905. His concepts of special and general relativity still provide our best descriptions of the large-scale Universe.

Relativistic Motion or speed comparable to the speed of light.

each other at speeds close to the speed of light itself. In such extreme situations, however, strange effects include objects growing shorter in the direction of their travel, and experiencing the flow of time more slowly (a phenomenon that needs to be taken into account in satellite navigation systems that rely on both fast-moving satellites and highly accurate atomic clocks).

Einstein's former tutor Hermann Minkowski (1864–1909) was among the first to embrace the relativity revolution, and also to develop an entirely new way of looking at the situation. He argued that we should treat time as another dimension intimately linked with the three space dimensions (and in some way 'at right angles' to them). The effects of special relativity can then be described in terms of a twisting or rotation of the dimensions around an object in relativistic motion, with the shortening of distance and extension of time effectively 'trading off' against one another.

C   This early example of a Minkowski diagram was drawn by Einstein's former tutor to explain the strange effects of special relativity. The blue-green lines show how the formerly perpendicular length (horizontal) and time (vertical) axes become 'tilted' as an object approaches the speed of light, resulting in time running more slowly and length contracting, as seen from the point of view of an outside observer.

D   A 'light cone' reduces the dimensions of space to a two-dimensional plane (S), with time (T) on the vertical axis. The upper cone (A), spreading out from the object, defines areas that the object's light and influence can eventually reach. The lower cone (B), coming to a point at an object's current location, defines the areas of spacetime that can be detected from, and therefore influence, the object.

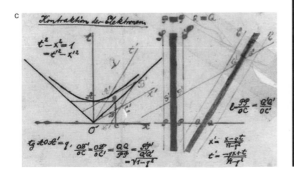

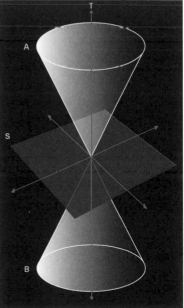

Minkowski's idea of a flexible four-dimensional 'spacetime manifold' in which dimensions can be warped, compressed and stretched proved hugely powerful, and Einstein put it at the heart of his general relativity theory of 1915. This theory was Einstein's model of gravity and acceleration, both of which, Einstein realized, are analogous to one another: it is impossible to distinguish the physics that occurs on a spaceship outside of a gravitational field but accelerated by its rocket motors from that on the same spaceship sitting still on a planet's surface and under the influence of gravity. Einstein realized the similarity is because gravity operates through the warping of spacetime; large masses effectively bend the dimensions of space slightly out of their ideal perpendicular arrangement, causing objects to fall towards them (extremely dense masses can also affect time).

This, then, is how space can have a shape: our Universe is full of matter, and that matter's mass can bend the lines of spacetime out of shape.

Over the past few decades, astronomers have seen this phenomenon at work, through a spectacular effect called gravitational lensing that influences the light of both individual stars and galaxy clusters 10 million light years across. Why, then, should it not also happen on the grandest scale of all, with the entire mass of the matter in the Universe bending the spacetime around it?

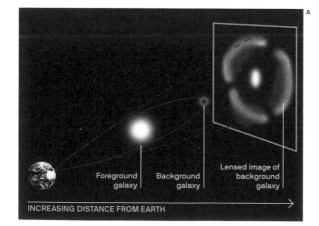

INCREASING DISTANCE FROM EARTH →

Foreground galaxy | Background galaxy | Lensed image of background galaxy

A   In gravitational lensing, light rays spreading out from a more distant object are deflected by warped spacetime around a closer massive object onto new paths that reach Earth. From an observer's point of view, the lensed object appears as a distorted ring around the foreground object.
B   This gallery of lensed galaxies was observed using the Hubble Space Telescope. Variations in the shape and intensity of lensing reveal the distribution of mass in the foreground objects.
C   These close-ups of gravitational lenses were captured by the Hubble Space Telescope's Faint Object Camera.

SDSS J1420+6019 | SDSS J2321-0939 | SDSS J1106+5228 | SDSS J1029+0420 | SDSS J1143-0144 | SDSS J0955+0101 | SDSS J0841+3824 | SDSS J0044+0113 | SDSS J1432+6317 | SDSS J1451-0239

SDSS J0959+0410 | SDSS J1032+5322 | SDSS J1443+0304 | SDSS J1218+0830 | SDSS J2238-0754 | SDSS J1538+5817 | SDSS J1134+6027 | SDSS J2303+1422 | SDSS J1103+5322 | SDSS J1531-0105

SDSS J0912+0029 | SDSS J1204+0358 | SDSS J1153+4612 | SDSS J2341+0000 | SDSS J1403+0006 | SDSS J0936+0913 | SDSS J1023+4230 | SDSS J0037-0942 | SDSS J1402+6321 | SDSS J0728+3835

SDSS J1627-0053 | SDSS J1205+4910 | SDSS J1142+1001 | SDSS J0946+1006 | SDSS J1251-1208 | SDSS J0029-0055 | SDSS J1636+4707 | SDSS J2300+0022 | SDSS J1250+0523 | SDSS J0959+4416

SDSS J0956+5100 | SDSS J0822+2652 | SDSS J1621+3931 | SDSS J1630+4520 | SDSS J1112+0826 | SDSS J0252+0039 | SDSS J1020+1122 | SDSS J1430+4105 | SDSS J1436-0000 | SDSS J0109+1500

SDSS J1416+5136 | SDSS J1100+5329 | SDSS J0737+3216 | SDSS J0216-0813 | SDSS J0935-0003 | SDSS J0330-0020 | SDSS J1525+3327 | SDSS J0903+4116 | SDSS J0008-0004 | SDSS J0157-0056

B

Gravitational lensing
The distorted images of distant objects created as their light rays pass through the warped spacetime around an intervening massive object that happens to lie along our line of sight.

The crucial issues, then, are just how much mass does it take to bend space, how does that compare to the amount of mass that is actually in the Universe, and what shapes might space bend into? Before we address these questions, however, there is one hugely important factor that needs to be taken into account: our Universe is not static, but expanding.

C

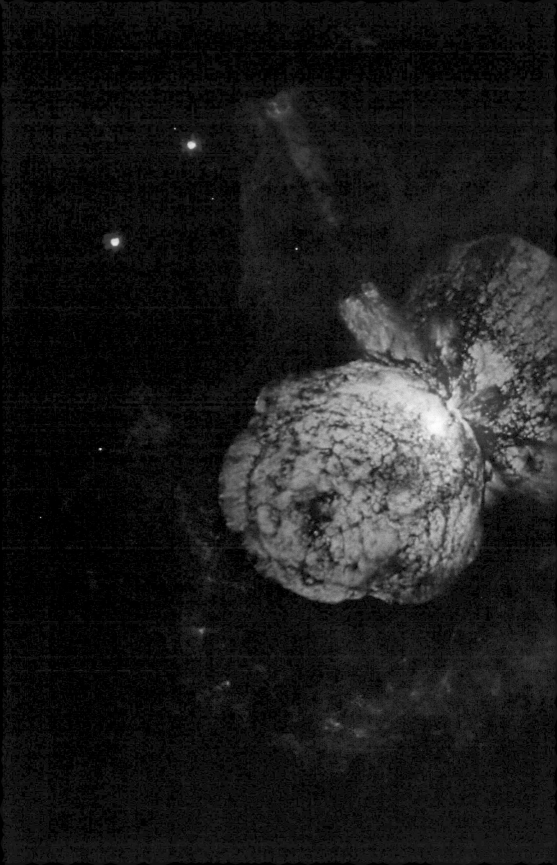

2. The Expanding Universe

A

The fact that our Universe is expanding completely changes the ground rules when it comes to addressing the shape of space.

Because the concentration of matter and mass is becoming less dense over time, its ability to warp space around it is, in most models, also diminishing. So how do we know that the Universe really is expanding, and why is it happening?

Vesto Slipher (1875–1969) This US astronomer pioneered the use of light spectra to measure the motions of galaxies, and to investigate the atmospheres and rotation of planets.

Isaac Newton (1643–1727) Today, Newton is best known for his laws of motion and universal gravitation that describe physical phenomena such as ballistics and planetary orbits, and for his formulation of the mathematical field of calculus.

Prism A wedge-shaped piece of optical glass that deflects different wavelengths of light onto different paths.

Spectrum (plural spectra) A band of light created by spreading out or diffracting light rays along slightly different paths depending on their wavelength and colour.

Wavelength The distance between successive peaks or troughs of a wave.

The first and most important piece of evidence for cosmic expansion was identified even before Edwin Hubble's confirmation in 1925 of the existence of galaxies beyond our own. As early as 1912, Vesto Slipher began a survey of the sky's mysterious 'spiral nebulae' from Flagstaff, Arizona.

Slipher analysed these distant objects using an optical device called a spectrograph, which splits the faint light gathered by the telescope onto slightly different paths depending on its wavelength or colour. The principle of splitting light in this way had been demonstrated in the 1660s by Isaac Newton, when he took a narrow beam of apparently white sunlight and passed it through a glass prism to create a rainbow-like spectrum.

Most natural light, including that from all stars, is similarly composed of many different colours (variations in the overall colour of objects are usually due to the distribution of the energy emitted or reflected over different wavelengths). However, the light from stars is so faint that, even when collected in a telescope, it was impossible to gather enough to create a visible stellar spectrum. It was only in the mid-19th century that this changed, with the invention of photographic plates that could absorb the faint smeared-out traces of light over many minutes or even hours in order to capture precious information.

A   Arp 274 appears to be a closely linked group of interacting galaxies, but in reality the central galaxy is moving away from Earth a great deal faster than the others, and is therefore much more distant.
B   Pages from Isaac Newton's *Opticks* (1704) explain key properties of light, including reflection, refraction and the presence of distinct colours within white light.

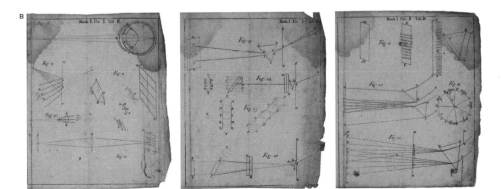

Slipher's spectrograph used the same basic principle as Newton's prism to create spectra from the light of faint spiral nebulae before capturing them in photographic form, and it led to some important discoveries. The fact the light of the nebulae formed star-like spectra at all was intriguing, because it suggested that the nebulae were not simply clouds of glowing interstellar gas (other more obviously gassy nebulae were known, but they emitted their light only in very specific wavelengths, similar to those released by burning elements in laboratory experiments). Instead, the spiral nebulae seemed to be formed of countless individual stars, presumably so far away that they were impossible to distinguish from a hazy mass.

Equally important, however, were the areas where the galaxies did *not* emit light. Just as in the spectra of the Sun and other stars, the light from the nebulae was missing at specific wavelengths, creating dark lines in certain parts of the spectrum.

Astronomers had realized in the mid-19th century that these lines were caused by atoms of gas in the upper atmospheres of stars absorbing light from below and, furthermore, had recognized that specific elements absorbed particular colours so that a star's 'absorption lines' could be used as chemical fingerprints to identify the elements it contained.

The discovery of spectral lines opened up a new means of analysing the way that a star was moving in space.

All stars are in motion, of course, but the 'proper' motion as they drift across the sky is imperceptible except over many years, and even then only measurable if the distance to the star is also precisely known through parallax measurements. However, if a star is in 'radial' motion, moving towards or away from us along our line of sight, there is another way of measuring even quite small rates of movement, known as the Doppler effect.

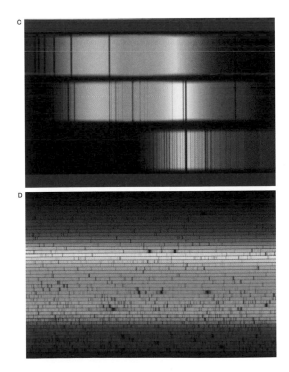

Nebula From the Latin for 'cloud', a nebula was the traditional astronomical term for any fuzzy, non-stellar object in the sky. Many objects once classed as nebulae are now known to be distant galaxies, and today the term tends to be used more strictly to describe clouds of interstellar gas and dust.

Atmosphere Stars are huge balls of gas all the way through, so in stellar terms, the atmosphere is the transparent outer layer overlying the visible luminous surface (itself known as the photosphere).

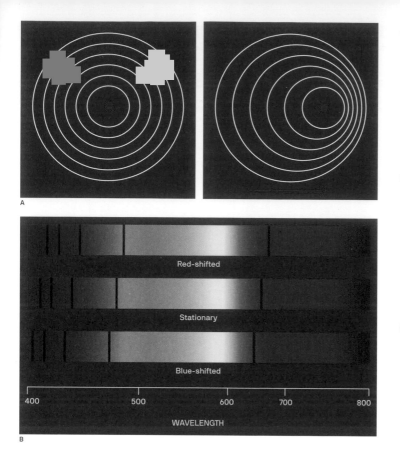

A One way of understanding Doppler shift is as follows: a static light source (left) emits light with wavefronts that spread evenly in all directions. If the source is moving, wavefronts are compressed (shortened) in the direction of motion (and spread or lengthened in the opposite direction).

B Although the difference in a star or galaxy's overall light output due to red and blue shifts is usually imperceptible and hard to quantify, the change in location of precisely defined absorption lines is easier to identify and measure, revealing the radial speed with which objects are moving relative to Earth.

C The unstable giant star Eta Carinae was long thought to be a single giant until tell-tale shifts in its spectral lines revealed that it is actually a tightly bound pair of binary stars, with masses of about 50 and more than 100 Suns.

Red-shifted

Stationary

Blue-shifted

| 400 | 500 | 600 | 700 | 800 |

WAVELENGTH

First predicted by Austrian physicist Christian Doppler (1803–53) in 1842, this effect is best known today from the shift in pitch we hear as an emergency vehicle rushes past us with its siren wailing. It is created because the waves emitted by a source (such as the sound waves from the siren) reach us with higher frequency (and consequently have shorter measured wavelengths) if the source is moving towards us, and in contrast have lower frequencies and longer wavelengths if the source is moving away. Despite the speed of light itself being constant, a similar effect alters the colour of starlight, so that when a star is moving towards us, its light appears bluer than normal (with a shorter wavelength) and if it is moving away, its light appears redder than normal (with its wavelength stretched out).

However, the Doppler effect is only useful if we have some means of knowing what colour a light source, such as a star, is supposed to be in the first place (the colour it would be if it had no radial motion whatsoever). This is where the absorption lines come in: although stars vary a great deal in colour, the precise pattern of dark lines associated with a particular element is locked to specific wavelengths. If that pattern is out of place, shifted towards the red or the blue end of the spectrum, it must be due to the Doppler effect.

The measurement of so-called blue shifts and red shifts in starlight was a huge leap forward in understanding the stars. It confirmed that many closely aligned stars in our sky are actually in orbit around each other, and also revealed 'wobbles' and splits in the lines of apparently single stars that showed they were actually closely bound binary pairs. Today, a more sophisticated version of the same basic technique has revealed that many stars are orbited by giant planets with enough mass to cause them to wobble very slightly.

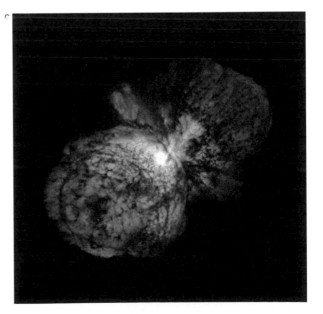

Binary A star system containing two members in orbit around each other. Most stars in our galaxy are members of binary or more complex multiple systems – as a singleton star, our Sun is in the minority.

A

B

For Slipher, however, the Doppler shifts of spiral nebulae presented something of a puzzle: not only were they generally larger than any shifts yet detected among stars, but also they were all in the same direction, towards the red end of the spectrum, indicating that the nebulae were moving away from Earth.

The high speeds of the nebulae convinced Slipher and others that they could not be associated with the Milky Way in the same way as normal stars. Where most stellar Doppler shifts indicated motion towards or away from Earth measured in metres per second at most, the nebulae were moving away at hundreds of kilometres per second – surely enough to escape the galaxy's gravitational pull?

Arthur Eddington (1882–1994) This British astrophysicist made many important breakthroughs in understanding the behaviour of stars. He was the first to successfully model the forces at work in their interiors, and to recognize the relationship between a star's mass and its energy output.

A   The appearance of the Andromeda Galaxy makes it easy to see how some astronomers suspected that spiral nebulae might be solar systems in the act of formation.
B   However, the bar-like central regions of many spiral nebulae were hard to explain as newborn Suns.
C   These two photographs of the solar eclipse in 1919 were captured during Arthur Eddington's expedition in order to measure the locations of stars near the Sun.

If these were stars or solar systems in the act of formation, why would they all be moving at such high speeds? And even if you could find an explanation for that, why were there no equivalent blue-shifted nebulae moving towards us at high speed?

c

Slipher argued that the red shifts pointed to spiral nebulae being separate systems far beyond our own galaxy, but it was to be some years before Hubble's confirmation. In the meantime, a revolution in our understanding of the nature of space took place, in the form of Einstein's theory of general relativity.

Published in a German scientific journal in the midst of World War I, Einstein's work was largely overlooked in the wider world until it was championed by the respected astrophysicist Arthur Eddington. In 1919, Eddington led an expedition to observe a total solar eclipse from the island of Principe off the west coast of Africa, where he measured shifts in the positions of stars close to the Sun – the 'gravitational lensing' effect described earlier. Many theoretical physicists now began to look in detail at the implications of Einstein's work, and to investigate one particular problem.

If, as general relativity suggested and gravitational lensing apparently proved, the presence of large masses warps space itself, then why was the Universe not collapsing on itself? At the time, astronomers were in general agreement that the Universe was both static in size and eternal in lifespan, so if relativity was right, gravity would surely have pulled all of matter, space and time together in a cataclysmic crunch long ago?

A

In order to solve this conundrum, Einstein added an extra term to his 'field equations' – the mathematical model that puts numbers on the relationships between gravity, mass and curving spacetime in general relativity. Known as the 'cosmological constant', it creates a constant negative pressure that counters the inward pull of gravity and ensures the scale of spacetime remains static. Einstein later called this idea his greatest mistake – he didn't live to see its surprising resurrection in the late 20th century as a possible explanation for the mysterious 'dark energy' we will encounter in Chapter 3.

As cosmologists began to grapple with the field equations, they discovered various interesting 'solutions' that could describe spacetime in special circumstances. One of these was the theoretical possibility that infinitely dense points called singularities could exist in Einstein's universe (the heart of the objects we now know as black holes). Another was the realization, first outlined by Alexander Friedmann in 1922, that the field equations allowed for an expanding Universe, and in this case you could do away with the cosmological constant altogether.

Black hole An infinitely dense concentration of mass with intense gravity that pulls in anything that strays too close and does not even allow light to escape.

Alexander Friedmann (1888–1925) This Russian mathematician and physicist developed equations that describe the shapes space can take in general relativity while still conforming to the observed characteristics of the Universe.

Georges Lemaître (1894–1966) Belgian astronomer Lemaître is generally regarded as the originator of the Big Bang theory, and predicted the relationship now known as Hubble's Law before Hubble himself discovered it. His support for an expanding Universe at first brought him into conflict with Einstein, who delivered the well-known rebuke: 'Your calculations are correct, but your physics is atrocious.'

Friedmann's discovery was treated as little more than a curiosity at the time (a golden rule of scientific theories is to keep them as simple as possible unless evidence emerges to demand a new approach, and Friedmann suggested no way of proving his idea). However, a few years later in 1927, Georges Lemaître returned to the topic with far greater success. He reached the same conclusion from Einstein's equations, and argued that there would be an observational consequence: the further away we look in space, the faster things would appear to be moving away from us.

This is not because our galaxy is somehow uniquely unpopular, but rather that galaxies are scattered across expanding space, rather like currants in a rising cake mix. Galaxy motion would then simply be a natural result of that general expansion; assuming that each light year or parsec of space is stretching at a uniform rate, then the further away we look, the greater the stretching effect that we would detect.

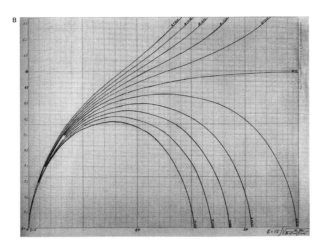

A   A computer simulation sequence shows how black holes distort spacetime around them, resulting in extraordinary gravitational lensing effects as light passes close by. The three images simulate a pair of stellar-mass black holes approaching and merging.

B   Georges Lemaître plotted various possible solutions to Einstein's field equations in his 1927 paper. Each corresponds with a different outcome for the evolution of spacetime and the future of the Universe itself.

One popular analogy is to imagine galaxies as dots on the surface of a balloon – as the balloon inflates, each dot moves away from all the others as the gaps between them expand, and the distance between the most widely spaced dots expands most rapidly.

This was exactly the effect that Hubble found when he followed up his initial studies of galaxies in 1929. Equipped with estimates of galaxy distances based on the brightness of their Cepheid variable stars, he compared these with new measurements of red shifts for the same galaxies obtained by his colleague Milton Humason (1891–1972). The resulting graph showed a clear, though noisy, relationship between galaxy distance and red shift.

Hubble's measurements showed only a very broad relationship; galaxy motions are also affected by the influence of gravity from other nearby galaxies, and this left plenty of room for different estimates of the precise rate at which the red shift increased with distance (a factor now known as the Hubble constant).

A   Milton Humason (far left) and Edwin Hubble (second left) pose with a visiting Einstein in the library of Mount Wilson Observatory in 1931.
B   Hubble's 1929 graph plots red shift against estimated distance for measured galaxies. Distances are given in megaparsecs (where 1 Mpc = 3.26 million light years).
C   By 1931, Hubble and Humason had extended their chart to incorporate far more distant galaxies.
D   This map of large-scale galaxy distribution across the nearby Universe is based on data from the Sloan Digital Sky Survey. In order to compile the map, astronomers used red shift as a direct indicator of distance. The numbers along the central axis indicate 'z', the ratio of the degree of red shift to the originally emitted wavelength of light.

M. L. Humason. Edwin Hubble   A. Einstein W. W. Campbell W. S. Adams

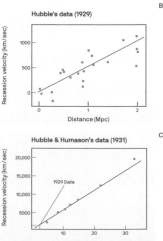

A

B
Hubble's data (1929)

Recession velocity (km/sec)

1000

500

0

0   1   2
Distance (Mpc)

C
Hubble & Humason's data (1931)

Recession velocity (km/sec)

20,000

15,000

10,000

5,000

1929 Data

10   20   30
Distance (Mpc)

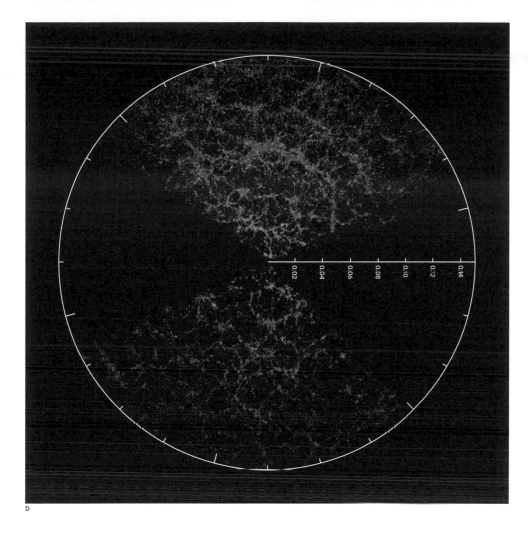

0.02 0.04 0.06 0.08 0.10 0.12 0.14

D

However, just as we saw with stellar spectra in Chapter 1, once a relationship has been established it is easy to apply it the other way around. Confident that red shift and distance are linked, astronomers can cast off the shackles of Hubble's variable-star method (inherently limited by the fact that we can only detect individual stars in the closest galaxies) and use red shift itself as a proxy for distance. It is not even necessary to calculate a distance figure in terms of light years; you can map the distribution of galaxies in terms of red shift alone (indicated by the letter 'z' in such equations).

This is the method that underlies today's large-scale maps of the Universe – highly sensitive spectrographs collecting the light of millions of galaxies and plotting their direction in space along with the red shift to produce three-dimensional charts of the way matter is distributed.

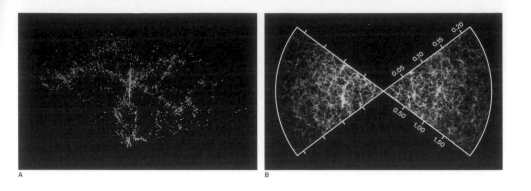

A

B

The first such charts, compiled in the 1970s, delivered an unexpected picture that has been reinforced by later surveys – matter in the Universe is clumpy. Recognition of the true nature of galaxies in the 1920s had gone hand in hand with the realization that such objects were so large and massive that their gravity draws them together in clusters of anything from a few dozen to more than a thousand members, but these surveys showed that the clusters themselves clumped together in huge superclusters.

A   Astronomers working at Harvard in the 1980s were the first to discover the uneven distribution of galaxies in large-scale structures called filaments and voids.

B   The Great Wall is a sheet of galaxy superclusters, 300 million light years away. With dimensions of about 200 by 600 million light years, it is one of the largest structures in the known Universe.

C   While 'wedge' maps show galaxies across just a narrow plane, this map from the Two Micron All-Sky Survey shows their distribution across the sky. (Note how the blue band of the Milky Way blocks our view in certain directions).

Galaxy cluster A group of galaxies bound together by gravity within a volume of space, typically about ten million light years across.

Supercluster A network of galaxies around a hundred million light years long, containing many separate galaxy clusters within it. Superclusters are held together by gravity, but their general shape and distribution are thought to arise from variations in the concentration of matter during the Big Bang itself.

What was more, even the superclusters aligned themselves into structures hundreds of millions of light years long, given evocative names such as the Stick Man and the Great Wall. The large-scale Universe consists of chains and sheets of galaxy superclusters, collectively known as filaments, surrounding vast and apparently empty regions called voids. We will see in Chapter 3 how this presented a significant challenge for astronomers to explain.

Surprisingly, making the final leap from the red shift pattern discovered by Hubble to general acceptance of Lemaître's expanding Universe proved far from straightforward. Hubble found that if he worked backwards from any plausible modern expansion rate to estimate how long ago everything might have been located at a single point in space, a potential cosmic origin point, the answers suggested a Universe that was 'only' two to three billion years old. And that could not be correct, since by this time geologists had already estimated the age of the Earth itself at more than 4 billion.

How could our planet be older than the Universe?

c

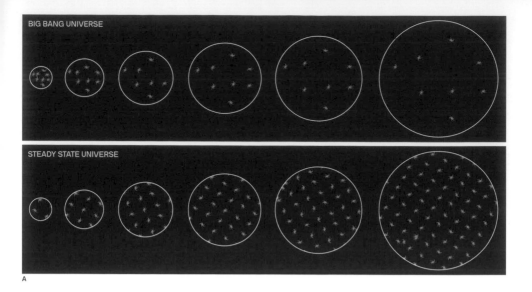

A

Hubble, therefore, expected that someone would find an alternative cause for the red shifts. The true explanation only became clear in the 1930s as astronomers established that there are actually two distinct types of Cepheid-like variables: the true Cepheids and the intrinsically fainter but otherwise similar RR Lyrae stars. Using the latter type in many of his measurements had led Hubble to exaggerate the Hubble constant.

Despite Hubble's doubts, Lemaître remained confident of his theory, and in 1931 made an important leap forward by tracing the expansion of the Universe backwards. Just as air heats up when compressed in a bicycle pump, he realized that the temperature of the Universe must have been much higher in its earlier denser state. Ultimately, he argued, the entire cosmos must have originated in a tiny, superhot, superdense 'primeval atom'.

Lemaître's ideas were met with opposition at first from an astronomical establishment that was attached to the idea of an eternal and essentially unchanging Universe. Friedmann, for example, accepted the idea of present expansion and a hot dense past, but suggested that perhaps the Universe was cyclic, going through alternating phases of expansion and contraction.

More popular were 'steady state' theories, ideas put forward by several leading cosmologists that new matter was created in some ongoing process, ensuring that cosmic density remained the same even as the Universe expanded. Ironically, it was one of the steady state's most ardent supporters, British astronomer Fred Hoyle (1915–2001), who during a talk in 1949 coined a new nickname that he intended to denigrate Lemaître's ideas, dismissing them as nothing but a 'Big Bang'.

By this time, however, two breakthroughs had already been made that would see the Big Bang rise from one of several competing cosmological models to the dominant explanation of conditions in the past, present and future Universe. In 1948, Ralph Alpher and Ukrainian-born American George Gamow (1904–68) had demonstrated how matter and energy would have been interchangeable in the early Universe (in line with Einstein's equation $E=mc^2$).

A In a Big Bang Universe (top), the gaps between galaxies get steadily larger as the space between them expands. In a steady state Universe (bottom), matter is created continuously so the density of galaxies should remain broadly even.
B Lemaître's ideas grew in popularity through the 1930s and 1940s. In a BBC radio interview broadcast in 1949, Fred Hoyle memorably described the theory as a 'Big Bang'.

Ralph Alpher (1921–2007) Alongside his doctoral supervisor Gamow, this US cosmologist developed the mathematical model to describe how energy released in the Big Bang transformed into elements. His work was considered so important that more than 300 people, including journalists, attended the presentation of his PhD dissertation.

$E=mc^2$ Einstein's statement that mass and energy are equivalent: the energy contained within a body is equal to its mass times the speed of light squared.

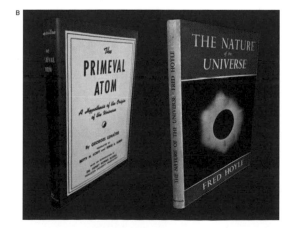
B

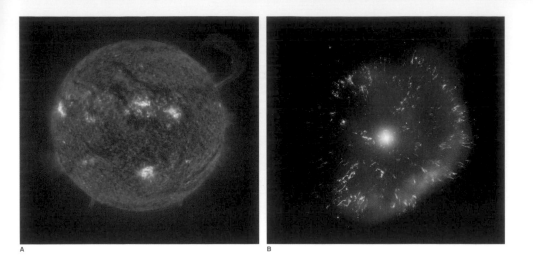

A

B

As space expanded and the temperature and energy density dropped, particles of matter would have coalesced out of the energy in proportions that accounted perfectly for the mix of elements in the basic ingredients of the cosmos. This resulted in vast amounts of the simple element hydrogen, smaller quantities of helium, the next simplest, and barely anything else (all the heavier elements found in today's Universe are a result of these raw materials being processed by nuclear fusion and other processes in the heart of stars).

The Big Bang, therefore, offered an important explanation for a specific feature of the Universe that could not be explained by its rivals.

A   Stars such as the Sun generate energy and shine by nuclear fusion, forcing together the nuclei of lighter elements to make heavier ones. They burn out when their fuel is exhausted.
B   At the end of a star's life, its outer layers are expelled into space, scattering elements across the cosmos for incorporation in later stars and planets. Violent supernovae that end the lives of the most massive stars can briefly push fusion to new limits, creating the rarest, heaviest elements.
C   This 'horn' radio antenna is of the type used by Arno Penzias and Robert Wilson in their discovery of Cosmic Microwave Background Radiation.

The second breakthrough, meanwhile, was in the form of a testable prediction. Also in 1948, Alpher and US scientist Robert Herman (1914–97) demonstrated that the expanding fireball of the early Universe would have left an afterglow that should still be detectable in today's cosmos. Radiation from this afterglow would be spread out across the entire sky, and red shifted more than anything seen so far – to a point where it would not appear as visible light at all, or even as infrared 'heat radiation'. Instead, it would take the form of short-wavelength radio waves called microwaves, heating the Universe to just a few degrees above absolute zero (the coldest possible temperature).

Alpher and Herman's prediction remained a curiosity until 1964, when it was discovered more or less by accident. Testing a highly sensitive new radio antenna at the Bell Telephone Laboratories in New Jersey, radio astronomers Arno Penzias (b. 1933) and Robert Wilson (b. 1936) found their observations seemed to be plagued by faint but continual 'noise'.

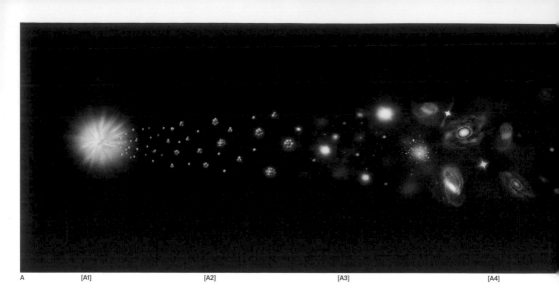

A   [A1]   [A2]   [A3]   [A4]

The signal seemed to come from all over the sky and could not be explained by any artificial sources (or even by pigeons that had taken up residence in the antenna). When they analysed it, they found it corresponded to a uniform 'background radiation' around 4°C (7°F) above absolute zero. The potential link with the predicted afterglow of the Big Bang did not occur to Penzias and Wilson until a colleague pointed out a recent paper from a team planning a deliberate search, but it soon became clear that this was exactly what they had found.

Today, we call this signal from the infancy of the Universe the Cosmic Microwave Background Radiation (CMBR). Measurements soon refined its average temperature to 2.7°C (4.9°F), and provided clinching evidence for the Big Bang theory – here was an observation that had not only been predicted before its discovery, but which rival cosmological models had no way of explaining.

[A5]　　　　　　[A6]　　　　　　[A7]　　　　　　[A8]　　　　　　　　　[A9]

Confirmation of the Big Bang refocused interest on the question of just how fast the Universe is expanding.

A The 13.8 billion years since the Big Bang have seen the raw materials of the Universe steadily enriched with more and more heavy and complex elements, giving rise to the elements incorporated in our own planet, Earth.

A1 Big Bang, 13.8 billion years ago.

A2 Subatomic particles coalesce into atomic nuclei and complete atoms by 400,000 years.

A3 After c. 200 million years, the first stars form.

A4 Remnants of the first stars form cores for galaxy formation.

A5 Galactic winds blow enriched material out into intergalactic space.

A6 Individual stars process light elements into heavier ones.

A7 Star death scatters heavy elements through galaxies.

A8 Supernova explosions create the heaviest elements of all.

A9 Heavy elements form rocky planets.

B The first map of the CMBR was produced by the COBE satellite in 1992.

C A far more detailed view was produced using nine years of observations by the WMAP satellite in 2012.

D This ultrafine CMBR map was produced by the European Planck satellite in 2013.

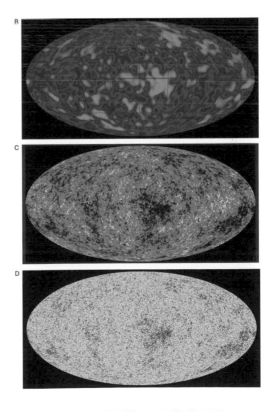

Assuming the Universe grew steadily since the Big Bang means that today's rate of expansion, as measured by the Hubble constant, also reveals how far back the expansion began, and therefore offers a proxy measure for the age of the Universe itself. Already in 1957, Allan Sandage had made vast improvements on Hubble's early estimates, taking account of the deceptive RR Lyrae stars and coming out with a value for expansion of 75 kilometres (46 mi) per second per megaparsec. This means that the speed at which galaxies are receding (once local variations are averaged out) increases by 75 kilometres (46 mi) per second for every 3.26 *million* light years of distance. So we would expect a galaxy 9.8 million light years away to be receding at a speed of around 225 kilometres (140 mi) per second, and one 98 million light years away to be retreating ten times faster.

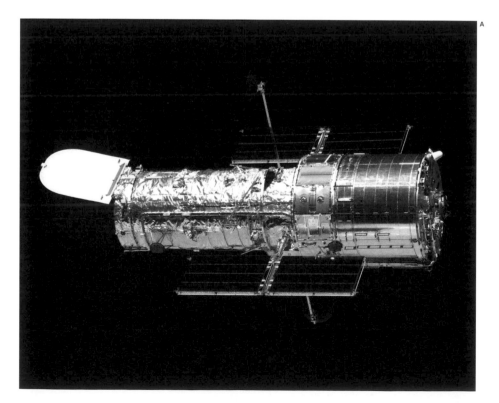

A

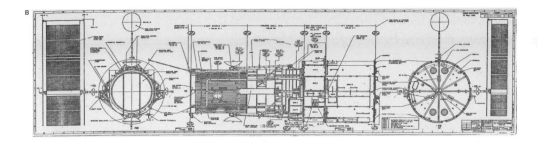

Sandage's figure for the Hubble constant was about 1/6th of the value estimated by Hubble himself, and despite being prone to a lot of uncertainty, it was still enough to pin down the age of the Universe at between 10 and 20 billion years. Over the following decades, many more attempts were made to refine the value, and pinning it down was the designated 'key project' for the groundbreaking Hubble Space Telescope (HST). Launched in 1990, the HST spent much of its first decade tracking Cepheid variables in distant galaxies, homing in on a final result (72 kilometres (44.7 mi) per second per megaparsec) that was pleasingly close to Sandage's early estimate. On this basis, cosmologists today estimate that the Big Bang occurred some 13.8 billion years ago.

A The HST's unique location above Earth's atmosphere gives it the clearest views of any optical telescope.

B As shown in these blueprints from 1981, the HST is 13.2 metres (43¼ ft) long with a main mirror 2.4 metres (7¾ ft) across.

C The HST was designed for periodic retrieval and servicing by the Space Shuttle. Repairs and upgrades have kept it at the cutting edge of astronomy for more than a quarter of a century since its launch in 1990.

Allan Sandage (1926–2010) US astronomer Sandage not only refined models for the age and expansion of the Universe, but also worked on one of the first models of the way galaxies form. He is credited with identifying the first quasar – a distant galaxy with an intense source of radiation in its core that gives it a deceptively star-like appearance.

Megaparsec A unit of astronomical distance measurement equivalent to 1 million parsecs or 3.26 million light years.

A

A This crowded galaxy cluster, known as MACS J0416.1-2403, is 4 billion light years away. It has probably changed beyond recognition since the light we currently see started its long journey to Earth.

B 'Deep-field' images of small areas of the sky can pick out even more distant galaxies, around 10 billion light years away. Seen in the early days of their formation, such galaxies often have active galactic nuclei.

The combination of the vast scale of space and the limited (though huge) speed of light turns the Universe into a natural time machine that can reveal a great deal about conditions in earlier times. For every 100 million light years we look into space, we are also effectively looking 100 million years back in time.

In the nearby Universe, this is merely a curiosity: it is interesting to know that we are seeing light that left a star many centuries ago, or a galaxy tens of millions of years ago, but on the enormous timescales of stellar and galactic evolution it makes little difference to what we actually see. Over larger distances, however, the cosmic time machine makes a significant difference.

Thanks to today's more powerful telescopes, we are now capable of detecting faint radiation from objects many billions of light years away, significantly closer to the early days of the Universe when everything was first coming together.

Galaxies in this earlier epoch appear less structured than today's elegant spirals. Collisions and mergers between them are more common, and they frequently have blazing energy sources known as active galactic nuclei at their hearts. These distant systems show the early stages in which galaxies came together, building up by collisions of smaller irregular star clouds to eventually form today's vast rotating discs, spirals and balls of stars.

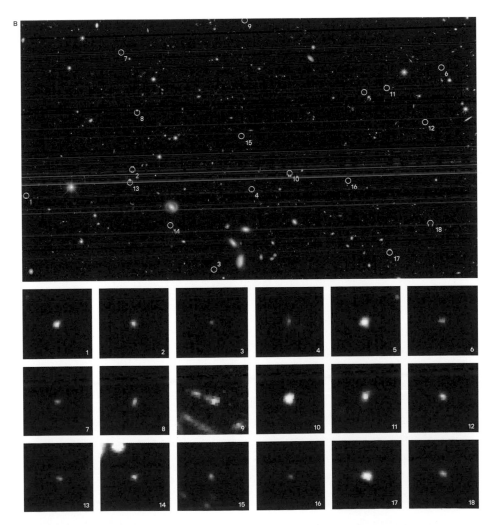

The most distant galaxies have truly enormous red shifts – the space between us and them is stretching so much that they are retreating at a significant fraction of the speed of light, and their light waves are stretched so much that they start to fade into invisible infrared radiation.

The HST's 'near-infrared' capabilities can reveal some galaxies too faint and distant to be seen in normal light, but beyond 10 billion light years or so, the red shift becomes so extreme that even the HST can see no further. Hence one of the key goals of NASA's successor to Hubble, the infrared James Webb Space Telescope, is to track galaxies even closer to the cosmic dawn.

However, there is one source of radiation that we can see from even further back in time – the faint glow of the CMBR.

A

A   The huge James Webb Space Telescope combines 18 hexagonal gold-plated beryllium segments for a total mirror diameter of 6.5 metres (21¼ ft). Designed for near-infrared astronomy, it will also detect red and orange visible light, including that from the most distant visible objects in the Universe.

B   The Planck probe reveals curious features of the CMBR (top), such as a slight difference in the average temperature of its two hemispheres, and a large cold spot (circled). Polarization (the 'orientation' of its radiation, bottom) also reveals clues to the early Universe.

Infrared Electromagnetic radiation with wavelengths between 700 nanometres (billionths of a metre) and 1 millimetre. Infrared is emitted by processes too weak to produce visible light, but can still have a significant heating effect.

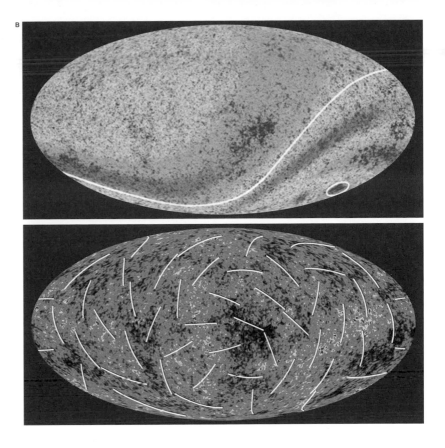

This radiation has travelled from the fireball of the Big Bang (or more accurately, the moment the initially foggy Universe became transparent, about 380,000 years later) across 13.8 billion light years of space to reach Earth. We see it in whichever direction we look, forming an impenetrable wall around the sky that is the closest we can get to viewing the Big Bang itself. As such, the CMBR has been a subject of intense study; tiny variations in the temperature of its radiation offer hints of the first structures emerging from the early Universe that blossomed into the large-scale concentrations of matter, the filaments and voids, present in the Universe today.

Due to its vast distance, meanwhile, the CMBR boundary is moving away from us at very close to the speed of light. The radiation that we now detect as feeble microwaves in the CMBR started out its journey as intense light as powerful as that from any star, before having its wavelength stretched as it crossed the vast gulf of expanding space.

This effect puts an ultimate limit on how much of the cosmos we can see – our observations of the Universe are limited to those objects whose light has had time to reach us in the 13.8 billion years since the Big Bang. In one way, then, the ancient Greek philosophers were right – we are the centre of our own Universe, an expanding bubble of spacetime whose ultimate boundary in every direction is defined by the time since the Big Bang.

However, this 'observable' Universe is not, as you might intuitively imagine, 13.8 billion light years in radius. It actually extends significantly further because we need to take into account the expansion of space during the most distant light's 13.8-billion-year journey towards us. This means that the spatial regions from which we currently see the CMBR emerging are 'now' about 46 billion light years away from us.

Of course, it is wrong to think of the observable Universe as the entirety of creation; every galaxy, every planet, even every person, is the centre of their own bubble of spacetime whose boundaries may stretch far beyond our own. 'Our' Universe might be finite, and an observer

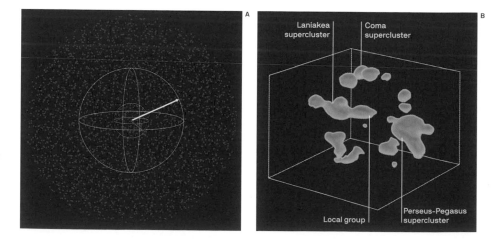

A

B

Laniakea supercluster

Coma supercluster

Local group

Perseus-Pegasus supercluster

A While we might expect our observable Universe to be limited in size to a 13.8-billion-light-year radius (pink), cosmic expansion means that the most distant objects whose light we can detect are now about 46 billion light years away (yellow), with the Universe itself extending far beyond that.
B The billion light years of space around Earth incorporates several enormous structures, including our own Laniakea supercluster, the Coma supercluster, and the Perseus-Pegasus filament.
C This unique view of the Universe maps the distance of cosmic objects on a logarithmic scale that increases by a factor of ten at each step.

c

currently sitting on a planet at its 'edge' would look in one direction and see the genesis of our part of the cosmos, but they could also look in the opposite direction and see regions forever beyond our view.

So does the expanding nature of the Universe mean that it goes on forever, an infinite sea of overlapping bubbles?

Friedmann's treatment of general relativity long ago showed that this depends fundamentally on the density of mass in the Universe, and how that bends space around it. Cosmic expansion is an important factor to take into account, but there is still no getting away from the question of mass, so that is where we turn our attention in Chapter 3.

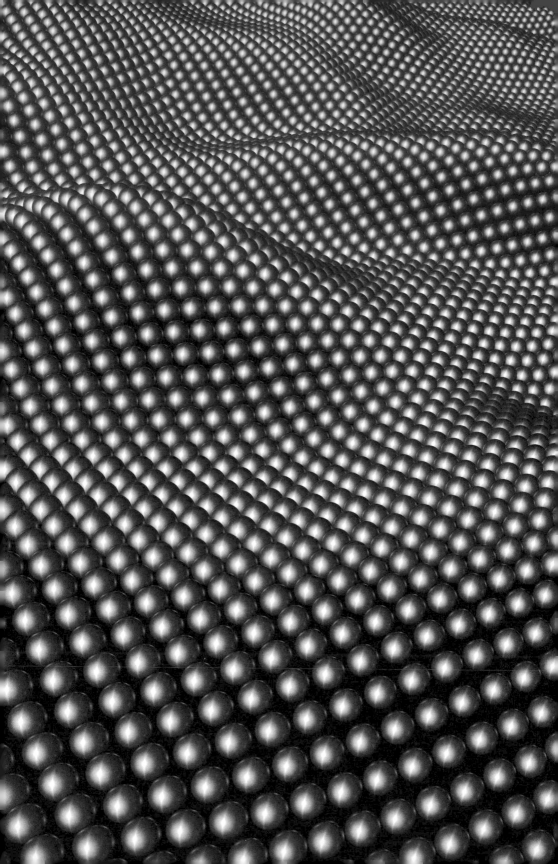

3. The Omega Factor

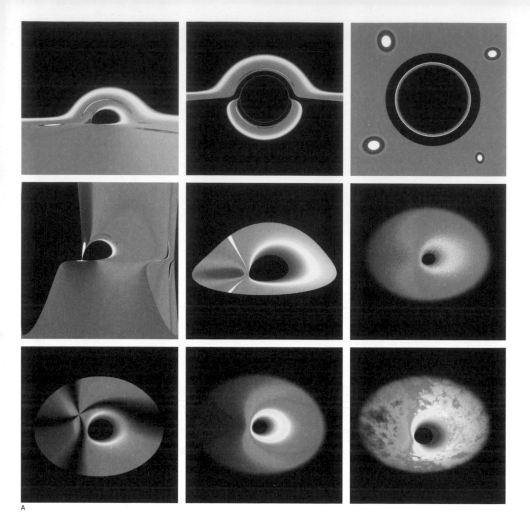

Questions about the Universe's shape always seem to lead back to this one deceptively simple-sounding question – how much mass does it contain?

A   A series of computer simulations of black holes model the effects of intense gravity warping spacetime on the three dimensions of space. Colours in these images reveal the amount of red shift – the stretching of light as it passes through warped space close to the black hole. The central black region, meanwhile, is the event horizon – the boundary beyond which even light cannot escape. Despite being a relatively small-scale phenomenon, models of black holes can reveal how the mass of the Universe itself, on the largest scale, can warp and distort the space it occupies.

In accordance with Einstein's theory of general relativity, mass warps and distorts spacetime, thus creating the effects we experience as gravity. The greater the mass and the denser its concentration, the more extreme the distortion and the stronger the gravity.

A common analogy to help visualize the distortions of general relativity is to discard one of the three space dimensions and imagine space instead as a two-dimensional membrane like a rubber sheet. Masses located on the sheet pull it out of shape, creating depressions known as gravitational wells, which deflect the motion of anything that passes too close. Black holes are an extreme example – a steep gravitational well around an infinitely dense point mass called a singularity, into which objects and radiation fall with no hope of escape.

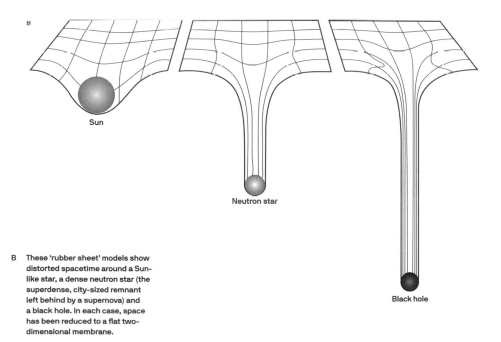

B

Sun

Neutron star

Black hole

B   These 'rubber sheet' models show distorted spacetime around a Sun-like star, a dense neutron star (the superdense, city-sized remnant left behind by a supernova) and a black hole. In each case, space has been reduced to a flat two-dimensional membrane.

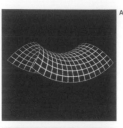

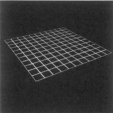

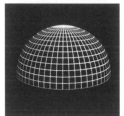

A   Top to bottom: an open,
    saddle-shaped universe,
    an open, flat universe,
    and a closed universe.
B   This graph shows possible
    evolutionary histories of
    the Universe related to
    different values of the
    density parameter omega.
C   Pages from Friedmann's
    paper of 1922, first outlining
    the possible overall
    curvatures of spacetime
    in general relativity.

If the mass of the entire Universe warps space in a similar way, then it gives rise to three broad possibilities. If there is sufficient mass, space could wrap around on itself like the surface of a sphere, creating a 'closed universe' with positive curvature, such that the paths of light rays that appear to be parallel in fact converge together over vast distances. Alternatively, if there is too little matter and the cosmic expansion triggered in the Big Bang dominates, then space could bend outwards in a saddle-like shape with negative curvature, so that parallel light rays would eventually diverge, a so-called 'open universe'. In between is a Goldilocks option where the amount of mass is 'just right', creating a scenario in which space remains flat, expanding in every direction but not curving inwards or outwards on the largest scales.

Fundamentally, these are the three possible 'shapes' of space on the largest scales, and as we will see later, these three scenarios also imply intriguing things about the eventual fate of our Universe.

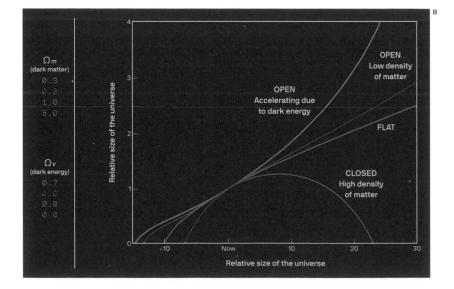

Ωm
(dark matter)
0.3
0.3
1.0
5.0

Ωv
(dark energy)
0.7
0.0
0.0
0.0

Relative size of the universe

OPEN
Low density
of matter

OPEN
Accelerating due
to dark energy

FLAT

CLOSED
High density
of matter

4

3

2

1

0

-10   Now   10   20   30

Relative size of the universe

c

In order to distinguish between open, closed and flat universes, therefore, we need to measure the mass of the Universe. But how can we do that?

Any attempt might at first seem doomed to failure simply because we don't know how far spacetime and matter might extend beyond the observable Universe that we can see. Fortunately, however, Alexander Friedmann's solutions to the field equations of general relativity (the mathematical model that describes a Universe of expanding spacetime similar to that in which we now know we live) give rise to a helpful shortcut.

The actual mathematics of the Friedmann equations is unsurprisingly abstruse, but an important consequence is that they describe how different characteristics of the Universe can arise depending on the values of a few key parameters. First and foremost among these is the observed density of the Universe – the amount of matter present *on average* in a unit volume of space. Whether the actual density is greater than a certain critical density affects the fundamental properties of space, and shapes the way the Universe will evolve from past to future.

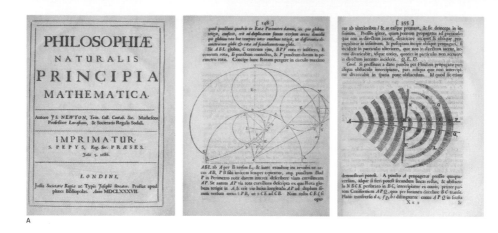

A

Cosmologists can combine these two terms into a simple number called the density parameter – the ratio of the observed density to the critical density, signified by Ω, the Greek letter omega. If Ω is greater than 1, then the actual density is greater than the critical density and the Universe is closed, whereas if Ω is less than 1, the observed density is less than critical and the Universe is open. A value for Ω of precisely 1 implies a Universe that is perfectly flat.

But how can we estimate the average density of something as fundamentally uneven as space appears to be? Fortunately, help is at hand in the form of a remarkable physical principle first outlined by Isaac Newton in his masterwork *Mathematical Principles of Natural Philosophy* (1687, commonly known as the *Principia*). Astronomers had spent much of the preceding two centuries in a more or less orderly retreat from the position that Earth was in a privileged position at the centre of the Universe with everything else circling around it. Newton was one of the first to realize that we could not assume the Sun's position to be any more privileged, and that we should guard against *any* tendency to believe we are seeing the Universe from a particularly special point of view.

Newton's cosmological principle, which became the bedrock of all subsequent theories about the Universe, is that the cosmos is both homogeneous and isotropic on the largest scale – in other words, it has the same average properties at all points in space, and it looks the same from all directions. Any conclusions we reach about the Universe from our particular point in space can therefore be applied to the Universe as a whole.

But if this is the case, how can we explain the apparent unevenness of the Universe at scales ranging from the distribution of mass in our solar system to the arrangements of galaxy superclusters? The answer is to go bigger still; there, on the very largest-scale maps of the Universe, covering scales of billions of light years, uniformity is finally waiting. The filaments and voids of the local Universe dwindle in scale and merge with countless other similar structures to form a distribution of matter that does indeed look the same at every position and in every direction, and that likely continues with little alteration far beyond the edge of our current observable Universe.

B

A   Isaac Newton's *Principia* (1687) lays down both fundamental laws of motion and the way in which these are modified by the presence of a resisting medium, before combining them into an overall system of universal gravitation.
B   This map showing some 2 million galaxies across a broad swathe of the southern sky reveals how the distribution of matter in the cosmos tends towards uniformity on the largest scales.

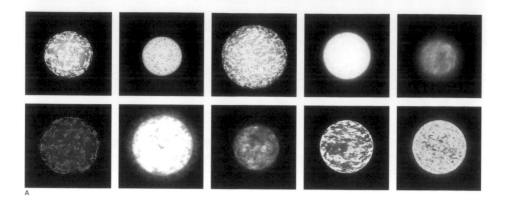

A

The average density of matter in the Universe as a whole, then, should be just the same as the density in those parts we can measure, provided we work it out on a sufficiently large scale. So how can we do that?

An important first step is to understand the true significance of stars. Not only are these vast balls of luminous gas the major concentrations of mass within their solar systems, but they are also pretty much the only objects capable of generating significant amounts of their own light.

Of course, not all stars are the same: they vary considerably in mass (from about 0.08 to tens of solar masses) and far more in overall energy output or luminosity (from about 1/100,000th of the Sun's output to a million times greater). Not all of this energy is released as visible light (stars with cooler surfaces emit mostly infrared radiation, whereas the hottest generate vast amounts of ultraviolet), but a century and a half of astrophysics has provided astronomers with the rules of thumb they need to estimate the likely mass of a star from clues such as the chemical fingerprints embedded in its spectral lines.

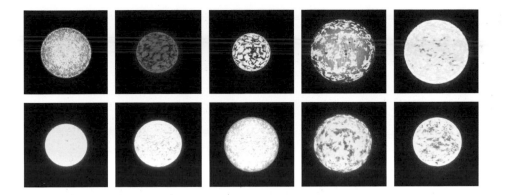

More of a problem is estimating the proportions of different types of star in our own galaxy, and getting some idea of whether our galaxy is typical of others in the Universe. Here, astronomers have to play a numbers game. Because a star's apparent magnitude falls rapidly with distance, highly luminous blue and white stars (and rare red giants) can appear bright over hundreds or even thousands of light years, while dim red dwarf stars can only be detected on our cosmic doorstep.

Apparent magnitude . The perceived brightness of an astronomical object as viewed by an observer. Apparent magnitude depends on both the object's true brightness (its luminosity or 'absolute magnitude') and its distance from the observer.

Red giant A star nearing the end of its life that becomes thousands of times more luminous than before due to changes in its internal structure, and also swells to 100 or more times its previous diameter. The red giant phase is relatively brief compared to the star's overall lifespan.

Red dwarf A low-mass star that shines very dimly and has a very low surface temperature, perhaps half that of the Sun.

The best way we can assess the number of these faint stars is to look for them in the neighbourhood of our solar system (often using infrared telescopes that are more sensitive to their feeble radiation) and then extrapolate. Based on this principle, it is clear that red dwarfs are far more common than bright stars. Of the 60 closest stars, 50 are red dwarfs (nearly all invisible to the naked eye). So while it might take several dwarf stars to equal the mass of one star like the Sun, their sheer abundance means that they probably account for more than half the weight of stars in our galaxy.

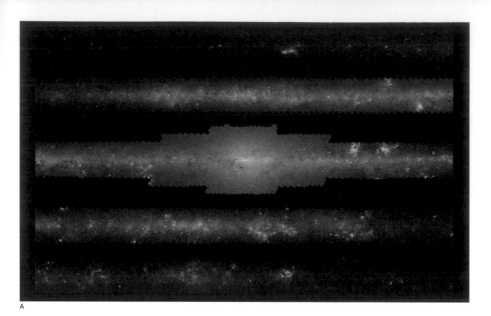

A

A   A series of infrared views looking towards the centre of the Milky Way reveals the presence of gas, dust and stars at different temperatures.
B/C Bright clouds of gas and dust called emission nebulae mark the sites where new stars are born within spiral galaxies. The most prominent nebulae in Earth's skies are the Carina Nebula (top) and the Orion Nebula (bottom).

Another factor we must take into account is the uneven distribution of stars in our galaxy and others. Spirals such as the Milky Way have a disc that is said to be 'dominated' by blue and white stars, often creating a distinctive spiral pattern. But this dominance is only in terms of which stars are generating most light – numerically they are still vastly outnumbered by the hordes of red dwarfs.

Close to the centre of a spiral galaxy, the rate of star formation trails off but the number and density of dwarf stars get much greater, so that they become the dominant source of light. Viewed from a distance, the sheer concentration of light from individually insignificant stars creates a bright hub region that can easily outshine the spiral arms.

The rules may vary slightly across different types of galaxy, such as the elliptiticals and irregulars, but in principle these techniques allow astronomers to estimate the number and distribution of stars within galaxies based on a combination of type, brightness and distance (either calculated using the Cepheid method described in Chapter 1, or estimated from red shift).

However, there is a lot more to galaxies than just stars – many are also filled with vast amounts of interstellar gas and dust, the raw materials from which new stars are formed, and into which they ultimately recycle most of their mass as they age and die. Estimates suggest that this so-called interstellar medium makes a moderate weight contribution in our galaxy at least, equal to perhaps 15% of the mass contained in stars. For the Milky Way, then, the overall mass of the stars and interstellar medium combined is thought to be between 50 and 70 billion solar masses.

Elliptical A ball-shaped galaxy containing large numbers of old red and yellow stars but little star-forming gas or dust. Ellipticals range from small diffuse clouds of tens of thousands of stars, to giants far larger than the Milky Way, containing up to a trillion stars.

Irregular A somewhat shapeless galaxy rich in gas, dust and new stars, usually smaller than the Milky Way.

Interstellar medium A mix of gas (predominantly hydrogen and helium formed in the Big Bang, with some heavier elements generated by stars during their lifetimes) and dust that lies between the stars.

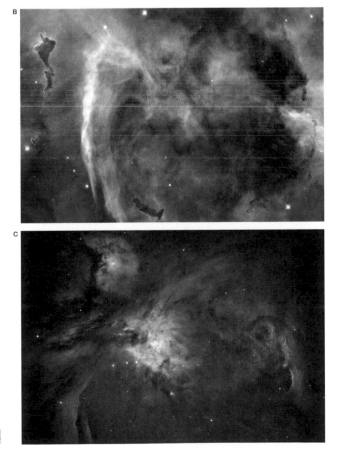

B

C

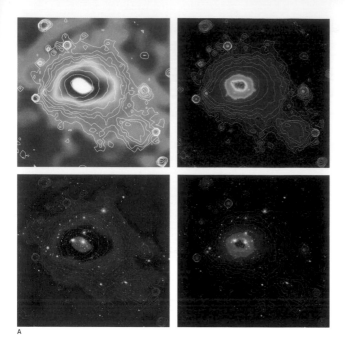

A

Fritz Zwicky (1898–1974)
Although this Swiss-American astronomer is best known for his discovery of dark matter in galaxy clusters, his other important contributions include the idea that supernova explosions are linked to the formation of superdense neutron stars, and his prediction that entire galaxies can act as gravitational lenses.

Centre of mass
The single point where the mass of a large object (or group of objects) can be considered as concentrated in order to simplify calculations of how it affects more distant objects.

A galaxy is even more than the sum of its stars, gas and dust.

Something else is also present: a mysterious 'dark matter' that outweighs the mass of everything else in a typical galaxy by a factor of about four to one. The name is deceptive; dark matter is not only dark (undetectable through emissions of electromagnetic radiation at any wavelength), it is also invisible and transparent. It simply does not interact with radiation *at all*, and can so far be detected only through the influence of its gravity.

The first hints of its existence came as early as 1933. Shortly after the confirmation of galaxies beyond our own, Fritz Zwicky realized it should be possible to measure the mass of galaxies based on how they affected their neighbours in close-packed galaxy clusters.

The principle is simple: the speed at which any object orbits around another much more massive object depends on both its distance from the central object and the amount of mass concentrated within it. For example, if the Sun suddenly doubled in mass, Earth would have to move along its orbit significantly faster to maintain the same distance from the Sun.

To keep things simple, Zwicky assumed that individual galaxies in a cluster are like planets orbiting a centre of mass where all of the cluster's mass effectively resides. When he put the idea into practice with a study of the Coma galaxy cluster, however, he found that individual galaxies were moving as if the cluster contained an astounding 400 times more mass than its starlight suggested.

A Images of the Coma galaxy cluster map the distribution of mass in the cluster from interactions with the CMBR (top left) and emission of X-rays (top right). The images below highlight the contrast between these density maps and the distribution of luminous matter in galaxies.

B According to NASA researcher Gary Prézeau, dark matter should concentrate in hair-like filaments around massive objects such as Earth (top) and Jupiter (bottom).

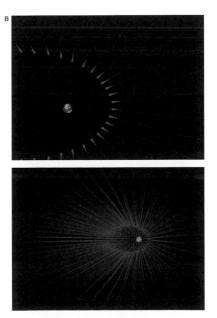

B

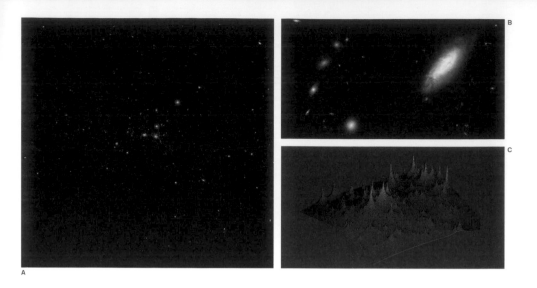

Zwicky named the mysterious unknown mass *dunkle Materie* or dark matter, but his discovery was largely ignored at first. After World War II, as the era of space-based astronomy got under way, astronomers assumed the missing mass was also likely to be explained by the huge clouds of previously undetected gas – sources of X-rays and radio waves – that lay at the centre of many galaxy clusters.

But while this 'cluster gas' (thought to be driven out of and to lose its attachment to individual galaxies as they age and evolve) accounted for much of Zwicky's missing mass, it turned out not to explain it all. In 1975, Vera Rubin confirmed that the same phenomenon is present much closer to home, inside the Milky Way itself.

Rubin was concerned with estimating our own galaxy's mass. Calculating the 'galactic rotation curve' (the speed at which stars orbit at different distances from the centre), she found that all the visible stars, interstellar dust and different types of gas could only account for about one sixth of the Milky Way's gravitational field.

Space-based astronomy The branch of astronomy that uses advanced technology, including satellites, to observe the Universe in ways that are impossible from the surface of the Earth, largely because of the way Earth's atmosphere blocks out radiations other than visible light and some radio waves.

Vera Rubin (1928–2016) Best known for her discovery of evidence for dark matter in the rotation of galaxies, US astronomer Rubin also courted controversy with claims that, once cosmic expansion is taken into account, galaxies are moving in peculiar directions across 100 million light years of nearby space. This eventually proved to be the first evidence for vast galaxy superclusters.

Baryonic matter Familiar forms of matter that are susceptible to the fundamental forces of nature including electromagnetism.

Gravitational microlensing A short-lived form of the gravitational lensing phenomenon, seen when light from a distant object such as a star is briefly distorted and brightened when a small dense object passes in front of it and distorts the path of light rays with its gravity.

A   The Coma cluster is a group of about 1,000 galaxies (mostly elliptical), some 320 million light years from Earth.
B   The motion of individual galaxies within the cluster suggested to Zwicky that they were influenced by far higher gravity than the cluster's visible matter suggested.
C   Today, astronomers can build maps of how dark matter is distributed, using various techniques to measure the effects of gravity in different parts of a galaxy cluster.
D   Vera Rubin and colleagues check their equipment at the Lowell Observatory in Flagstaff, Arizona, in 1965.
E   A composite of X-ray data (purple) and visible light (yellow) shows the concentration of hot gas in and around galaxy cluster Abell 1689.

When other astronomers backed up Rubin's discovery in the late 1970s, the search for dark matter began in earnest. Broadly speaking, explanations fell into two camps. One was that the missing mass was explained by compact, faint or dark objects composed of normal or baryonic matter – so-called Massive Compact Halo Objects (nicknamed MACHOs) that loitered in the halo regions above and below the discs of galaxies. Candidates included rogue planets, black holes and burnt-out, faded dwarf stars, but surveys aimed at sampling this population using gravitational microlensing have suggested that such objects are rare and contribute only a negligible amount to the overall missing mass.

D
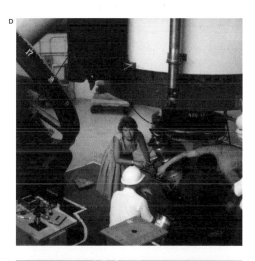

E

A

The other option is exotic matter:
new families of particles that
have mass and are susceptible to
gravity, yet are somehow immune
to interacting with electromagnetic
radiation. These Weakly Interacting
Massive Particles (known as WIMPs)
are today's leading candidate
for dark matter, even though they
remain essentially hypothetical.
The most sophisticated evidence
for their existence comes from
maps of the way in which apparent
clouds of these particles interact
with each other to create patterns
of distortion in the gravitational
lensing of distant galaxy clusters.

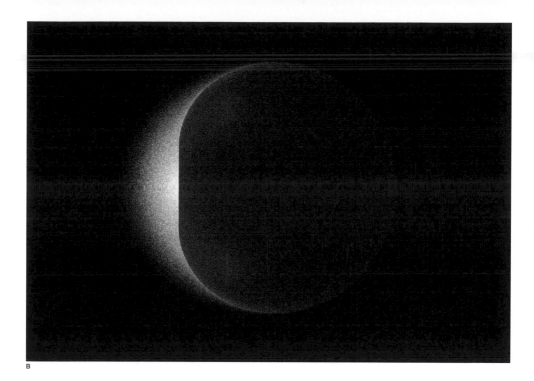

B

Yet even when the vast amounts of dark matter are taken into account, estimates of the density of the Universe turn out to be well below the critical density, producing values of Ω well below 1.

So can we say that the Universe is definitely open?

Fortunately, there is another way of looking at the problem, providing an independent cross-check of the ballpark figures gleaned from observations of today's Universe. It lies hidden in the Cosmic Microwave Background Radiation (CMBR) that we encountered in Chapter 2 – the radiation from the dawn of the Universe that is the most distant thing we can see.

A Dwarf galaxies such as the Fornax Dwarf are clouds of sparsely scattered stars, thought to be held together by dominant dark matter.

B A computer model shows how high-energy gamma rays should be produced as dark matter falls towards the event horizon of a black hole. Such emissions could offer a way of learning more about this mysterious matter.

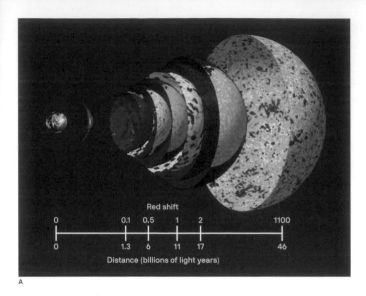

Red shift

0	0.1	0.5	1	2		1100
0	1.3	6	11	17		46

Distance (billions of light years)

A

The CMBR marks a very specific moment in the history of the young Universe: the point at which it became transparent. Directly after the Big Bang, space was filled with a blizzard of subatomic particles that rendered it opaque; rather like sunlight scattered in a fog bank, photons of radiation could only travel for short distances before interacting with matter particles and ricocheting in another direction.

However, as the Universe expanded and matter was less densely packed, it also cooled. Within minutes of the Big Bang, temperatures had fallen low enough for protons and neutrons to combine, forming the simplest atomic nuclei, but vast numbers of electrons still remained far too energized to be trapped into orbit around them and create true atoms. It was only after 380,000 years, as cosmic temperatures dropped below about 3,000 Kelvin, that nuclei were at last able to 'capture' electrons (a process known as recombination). The density of particles abruptly fell away, and radiation was, at last, able to travel in straight lines, racing away in all directions to create the CMBR.

Once you know where the CMBR came from, it is easier to understand why it might tell us something useful about the density parameter. The rate at which the cosmic temperature fell and the moment at which decoupling occurred are both indicators of how tightly matter was packed into the early Universe.

In addition, the presence of ripples in the CMBR – tiny variations in its temperature indicating slight differences in the overall density – can tell us even more. All that ricocheting radiation in the early Universe exerted an outward pressure that overwhelmed the attraction of gravity (and so prevented normal matter from clumping together), but dark matter was immune to the same effects, so it immediately began to coalesce around slight concentrations in the distribution of matter created in the Big Bang itself.

The intensity of the ripples, therefore, provides a measure of the balance between the two types of matter, and measurements from dedicated satellites such as NASA's Wilkinson Microwave Anisotropy Probe (WMAP) suggest that the Universe contains about 2.8×10^{-30} grams of matter per cubic metre of present-day space (0.00000000 0000000000000000000028 g/m^3). That figure, equivalent to about 1.7 atoms of hydrogen per cubic metre (or roughly 1 atom in every 20 cubic feet), tallies well with 'top-down' estimates of matter density from observations of today's cosmos, and should mean that Ω is well below 1 and the Universe is therefore distinctly open.

Case closed, you might think, but there's a problem. The CMBR also offers *another* independent way to directly double-check the density parameter, and this delivers a surprising and contradictory result.

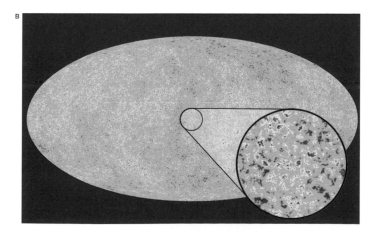

A   These nested spheres show the density of matter in the Universe at various distances and red shifts. They reveal a correspondence between the distribution of galaxies in the nearby Universe, and the structure of the CMBR (the outermost shell).
B   The WMAP's map of the CMBR shows cool, high-density areas in blue and hot, low-density ones in red.

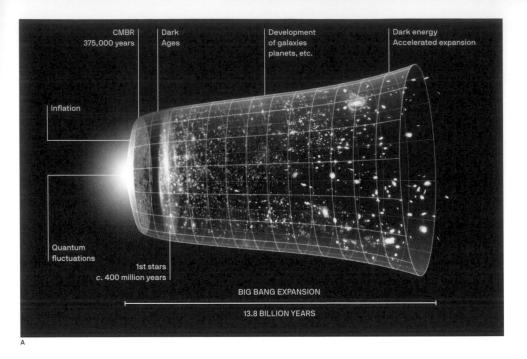

The idea is a simple matter of geometry.

As discussed previously, the different curvatures of space arising from different possible values of Ω cause light rays to converge or diverge over long distances, and this means that they should affect the apparent size of extremely distant objects.

Positive curvature would mean that light rays are converging and therefore would make distant objects appear larger than otherwise expected; negative curvature and diverging light rays would make them appear smaller. The obvious question, though, is what objects of predictable size can we see over sufficiently huge distances for the effect to be measurable?

The answer, happily, lies in the most distant structures we *can* see: the ripples of the CMBR itself. Other aspects of the CMBR allow cosmologists to predict the size to which structures could have grown before the recombination era, and therefore the angular size at which the brightest microwave ripples *should* appear in the sky (roughly one degree across, or about twice the size of the Full Moon). Scientists take the fact that the ripples do indeed appear at exactly the predicted size as a sign that their light rays have remained parallel in their journey across 46 billion light years of space. They can, therefore, independently show that, within the limits of current measurement technology, the Universe is geometrically flat.

But hold on – a flat Universe implies a density parameter Ω of precisely 1, and yet baryonic and dark matter combined are not nearly massive enough to produce this. What's going on? The strange answer to this question has only become clear in the past two decades, and it turns much of what we thought we knew about cosmology on its head.

A   Most cosmologists agree that the Universe expanded rapidly shortly after the Big Bang (an event called Inflation – see page 102), and has grown more steadily since. However, the discovery of dark energy (see page 88) suggests that its rate of expansion is now accelerating rather than slowing down.

B   Possible shapes of space and their effects on light rays: from top to bottom, positive curvature (converging rays), flat spacetime (parallel rays) and negative curvature (diverging rays).

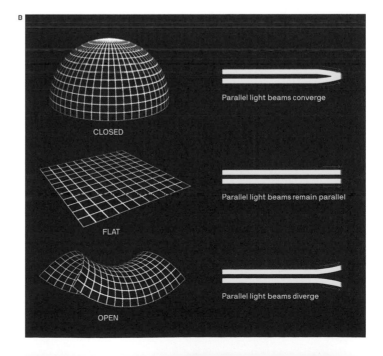

B

CLOSED

Parallel light beams converge

FLAT

Parallel light beams remain parallel

OPEN

Parallel light beams diverge

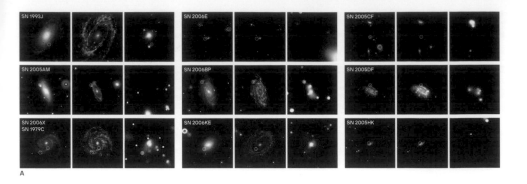

A

In the mid-1990s, astronomers keen to find an independent means of testing the cosmic expansion data being gathered by the Hubble Space Telescope hit upon an ingenious alternative to the classic 'Cepheid variable' method of measuring distances. Theoretical models suggested that a certain type of exploding star called a Type Ia supernova always released exactly the same amount of energy and therefore reached the same peak luminosity. This would, in theory, allow astronomers to use it as a 'standard candle' – an object whose intrinsic luminosity could be compared with its apparent brightness in Earth's skies in order to calculate its true distance.

There was one significant problem: Type Ia supernovae are rare one-off events. A galaxy such as the Milky Way sees perhaps one in every thousand years, and even that one bursts into brightness and declines back into obscurity in only a matter of weeks. Detecting a significant number of Type Ia supernovae in a short timescale therefore requires the long-term study of thousands of individual galaxies, something that only became possible with the rise of computerized astronomy. Furthermore, you are more likely to find the right type of supernova over great distances (where there are more galaxies to choose from) than in our cosmic backyard.

Type Ia supernova
A stellar explosion triggered by a neutron star that becomes too massive to support its own weight and therefore undergoes a sudden collapse into a black hole. Because this always happens at a particular threshold mass, the explosion always releases the same amount of energy and reaches the same peak luminosity.

Computerized astronomy The large-scale use of computers, electronic CCD detectors and other devices to automate the capture and processing of data, allowing astronomers to measure hundreds of objects in a single telescope observation rather than studying them one at a time.

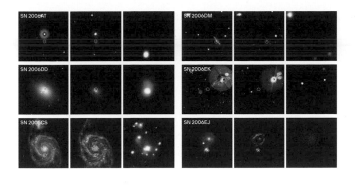

A This gallery of supernova explosions is captured in optical, ultraviolet and X-ray radiation. Supernovae can outshine entire galaxies and (if they are of the right type) provide a handy cosmic distance scale.

B The star V445 Puppis, a tight binary system currently shedding material in two 'bipolar' shells, is a likely future candidate for a Type Ia supernova.

Nevertheless, this makes 'supernova cosmology' a great way of checking cosmic expansion over large distances, and two teams – the international High-Z Supernova Search Team and the California-based Supernova Cosmology Project – set out to test the theory. Between then, they harvested data for 42 supernovae with high red shifts, implying distances of several billion light years, and 18 more in the relatively nearby Universe.

Because supernovae can be seen over much greater distances than Cepheid variables, the measurements stretched far beyond those made by the Hubble Space Telescope's 'Key Project'. The astronomers expected, therefore, that the more distant supernovae would be somewhat brighter than one might expect from a raw combination of their red shifts and the Hubble constant; cosmic expansion, they assumed, must have slowed down at least a little since the Big Bang, so the supernovae would be slightly closer than predicted.

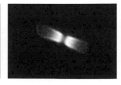

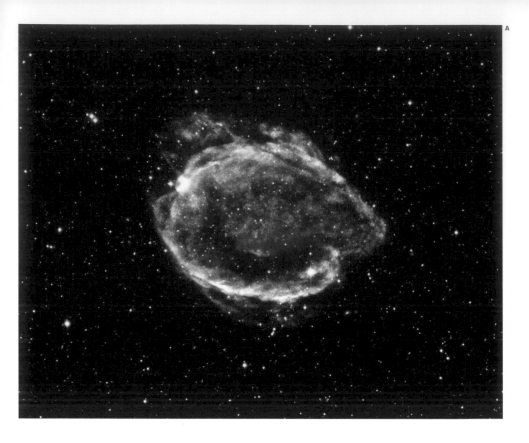

A

It was a huge surprise, therefore, when the exact opposite proved to be the case. The distant supernovae were consistently fainter than their red shifts predicted. In 1998, after months spent attempting to find other explanations, the astronomers published their evidence that cosmic expansion is, in fact, *accelerating*.

US cosmologist Michael Turner (b. 1949) soon coined the term 'dark energy' to describe the driving force behind the expansion, although the name itself does no more to explain what this mysterious force actually *is*. Two decades later, we are little wiser on this basic question, but most cosmologists nevertheless agree that dark energy is a real phenomenon, confirmed by subsequent observations that have not only pinned down the rate of acceleration, but also shown that it has changed over time.

Until about 7 billion years ago, it seems that the expansion of the Universe was indeed slowing down, before dark energy overwhelmed the steady deceleration caused by gravity and the rate of expansion began to increase.

For our purposes, the exact nature of dark energy is not as important as the implications of the simple fact that it exists, but it is still worth briefly explaining the rival theories that hope to describe it.

The first of these, the cosmological constant theory, revisits an idea first suggested by Einstein in 1915, that space has some inherent quality that causes it to expand over time, above and beyond the expansion triggered in the Big Bang itself. This quality is effectively a tiny amount of energy intrinsic to a fixed volume of space, which somehow creates a repulsive effect countering the inward pull of gravity. Undetectable in local space, the effect only becomes clear over great distances, and it also increases over time (as the volume of space in the Universe increases), matching neatly with the evidence that dark energy has strengthened over time.

A   A colour-coded image from the Chandra X-ray telescope reveals hot gas expanding in the so-called remnant of a Type Ia supernova. This particular explosion is thought to have been unusually 'lopsided'.

B   This graph shows cosmic evolution in three possible cosmologies – the traditional Einsteinian relativity (green), a Universe boosted by dark energy (red) and a newly proposed scheme (blue) that may explain cosmic acceleration without the need for dark energy. Each line on the graph corresponds to a predicted pattern of cosmic structure (left-hand column).

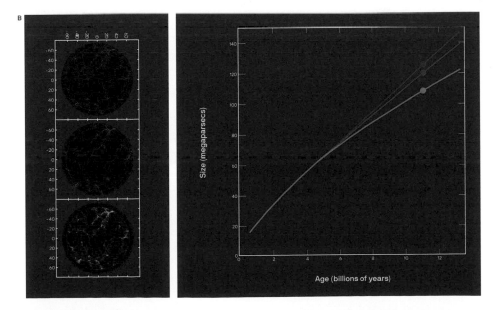

The major alternative explanations are so-called 'quintessence' theories, distinguished from the cosmological constant because they see dark energy as a non-uniform property, something that accumulates in certain regions of space and causes them to expand more than others. Quintessence theories are quite varied, but they all share a basic approach of treating dark energy as a fifth fundamental force, somewhat akin to gravity, electromagnetism and the forces within the atomic nucleus.

Whatever the nature of dark energy, it does provide a somewhat surprising explanation for the observed flatness of our Universe. Despite the fact that it is driving expansion rather than contributing to the inward pull of gravity, dark energy is nevertheless a contributor to the overall energy of space. It can, therefore, be considered as also having mass (through Einstein's $E=mc^2$ equation), and therefore shifting the density parameter Ω towards 1.

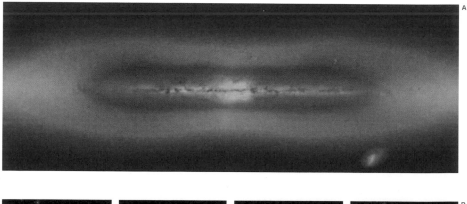

A

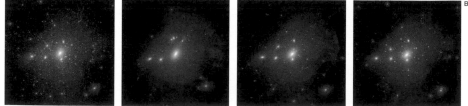

B

c

Measurements of the CMBR from WMAP and other satellites conclude that the mass/energy content of the Universe is a mere 4.9% ordinary baryonic matter, 26.8% dark matter and an amazing 68.3% dark energy. At this level, then, the Universe appears to be flat (parallel lines are really parallel, and at the largest scales the three-dimensional grid of space continues far beyond the limits of the observable Universe with its dimensions remaining orthogonal to each other. But despite this, the weird nature of dark energy means that it is growing larger at an ever-accelerating rate.

Two key questions remain: is this the *only* level on which we can interpret the shape of our cosmos, and what does it all mean for the future of the Universe?

Orthogonal Directions are said to be orthogonal if they are at right angles (precisely 90 degrees) to each other.

A   A composite image of the dark matter disc (red contours) and the Atlas image mosaic of the Milky Way, obtained as part of the Two Micron All Sky Survey (2MASS), by J. Read and O. Agertz.

B   These density maps predict the concentration of mass in the halo of a spiral galaxy according to four different dark matter models.

C   This conceptual artwork shows how our view of the Universe might be affected by the distorted spacetime approaching a black hole.

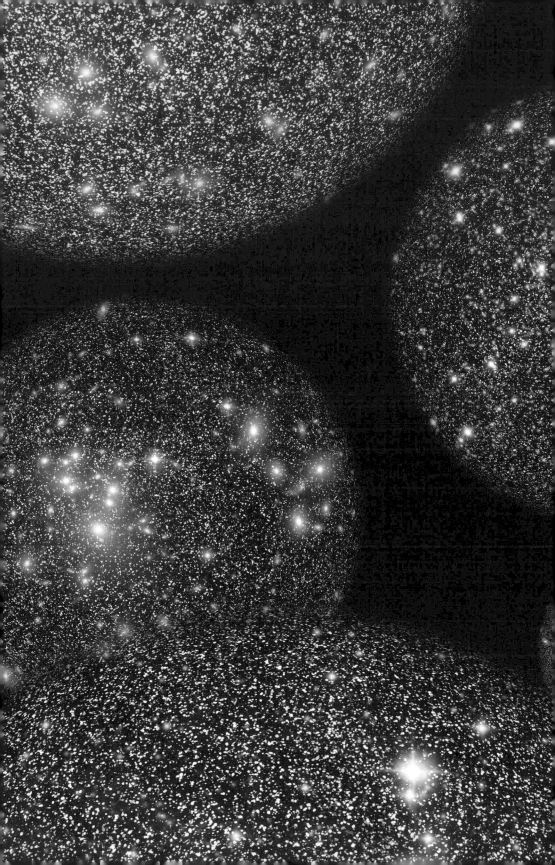

4. The Shape of the Multiverse

As we saw in Chapter 3, the recent discovery of dark energy (whatever it ultimately turns out to be) neatly explains the otherwise puzzling evidence from the microwave background that spacetime is 'flat' on even the largest scales, while also suggesting that the Universe is expanding at an accelerating rate.

We can, therefore, be sure that the observable Universe (the limited volume of spacetime we can observe thanks to the fixed speed of light) is an expanding spherical bubble roughly 46 billion light years in radius, across which the three perpendicular dimensions of space form a uniform 'grid'. Within this region, sometimes known as the Hubble volume, gravitational fields created by massive objects from planets to galaxy superclusters 'pinch' the dimensions of space together in local regions (simultaneously 'stretching' the time dimension to create an effect called time dilation), but these distortions pale into insignificance on the largest scales.

But what of the region beyond the observable Universe? Is there anything we can say about the shape of space *beyond* the arbitrary barrier created by the age of the Universe and the speed of light?

Hubble volume The spherical volume of space around any point in the Universe, defined by the distance light has been able to travel since the recombination era shortly after the Big Bang.

Time dilation A slowing-down in the relative passage of time for objects moving at high speeds. Time dilation is predicted by Einstein's theory of special relativity, and has been demonstrated to affect precise clocks flown aboard jet aircraft and navigation satellites.

A   Long-exposure images from powerful telescopes can capture galaxies to the limit of the observable Universe – but multitudes more lie forever beyond their grasp.
B   Scientists Joseph Hafele and Richard Keating demonstrated the strange phenomenon of time dilation in 1971 when they accompanied four atomic clocks on a round-the-world airline trip.

B

The most obvious conclusion is that space is infinite (in the simplest sense of the word) – we could keep on travelling in one direction forever and never run out of Universe to explore, or find ourselves back where we started. Based on current evidence, the outer barrier of our observable Universe is retreating from us at an accelerating rate that would prevent us ever reaching it, even if we could travel at the speed of light itself.

But more than this, even if we could somehow achieve the impossible and travel many times *faster* than the speed of light, we would still find ourselves with an infinite volume of space to explore. Our own personal observable Universe would, of course, move with us (since by definition, we are at the centre of it), so in Earth-centred terms if we reached the 'edge' of the observable Universe (a spot about 46 billion light

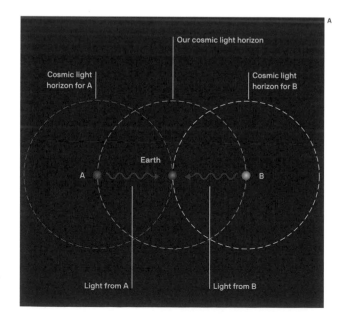

A

Our cosmic light horizon

Cosmic light horizon for A

Cosmic light horizon for B

Earth

A

B

Light from A

Light from B

Infinity In mathematical terms, any quantity that can be measured forever with no limit in a particular direction or sense. The simplest infinity is that of the 'real' or counting numbers, which can be counted upwards from 1 without end.

years away that has just been reached by light
released at the recombination era in the current
location of the Milky Way), we would be able
to look in the opposite direction into regions
of space forever hidden from Earth astronomers.

In fact, as we raced on across space, we would find ourselves
crossing an infinite number of Hubble volumes – overlapping
bubbles of spacetime like our own, each blown up from its own
infinitesimally small part of the primordial Big Bang and stretching
away in every direction. Therefore, it might seem reasonable to
imagine space on the largest scale as a vast sphere growing at
a tremendous rate (far faster than the speed of light) composed
of an infinite number of these rapidly expanding bubbles.

Case closed; that settles the question
of the shape of space. But not so fast.
This endless expanse of an apparently
'flat' Universe turns out to be just one of
several different types of cosmological
infinity, some of which imply the presence
of further dimensions beyond those
we normally perceive. And even if space
is flat in our familiar three dimensions,
then it could still take on bizarre shapes
in these higher dimensions.

Brace yourself, because things are about to get seriously weird.

Physicists generally refer to the complex structure containing an assembly of individual 'universes' (however you choose to define those) as a 'multiverse'. The term was first coined by US philosopher William James (1842–1910) in 1895 (although he was talking rather more loosely about ideas of perception beyond the mundane). The first person to suggest the multiverse might be a physical reality, meanwhile, was the renowned Austrian physicist Erwin Schrödinger, who floated the idea in an influential lecture in 1952.

The endless expanse of spacetime described above, with its infinite number of observable universes, is the simplest type of multiverse we can imagine. What can we deduce about its properties? Woven from the same cloth as the Universe we can see around us, and born from Big Bang conditions that we assume were the same everywhere, it is effectively just 'more spacetime'. If we could step into another part of the multiverse through a magic door, we might reasonably expect the basic conditions to be fairly similar: three dimensions of space and one of time, and physical constants operating in familiar ways.

A–C Particle physicists probe quantum behaviour by colliding subatomic particles at close to the speed of light in machines such as the Large Hadron Collider. Such collisions produce huge amounts of energy, briefly permitting the independent existence of particles that are otherwise confined within the atomic nucleus, such as the Higgs Boson (simulated in B and C).

D Antimatter is normally destroyed instantaneously on contact with normal matter, but it can be confined within magnetic fields. In 2011, scientists discovered a reservoir of 'antiprotons' (pink) trapped by our planet's magnetosphere just a few hundred kilometres above Earth's surface.

D

There might be some intriguing variations, of course: for example, although we cannot be sure, some parts of the multiverse might be dominated by antimatter.

Matter and antimatter were theoretically created in identical amounts during the Big Bang itself, and the dominance of matter in today's Universe remains something of a mystery. There are suspicions that it has something to do with a fundamental breaking of 'symmetry' in the interactions of certain subatomic particles, which would suggest the entire multiverse is biased in the same direction, but we cannot yet be entirely sure.

Erwin Schrödinger (1887–1961) Schrödinger was a key figure in quantum mechanics, developing equations to describe the way that particles can exhibit wave-like properties with uncertain values. He also played a leading role in debates about what this meant for physics.

Antimatter Particles with equal but opposite electrical charges to those of everyday matter. They have a nasty habit of annihilating completely on contact with normal matter particles, releasing an intense burst of energy.

Symmetry In the language of particle physics, a symmetrical interaction is simply one that behaves the same when some of its properties are inverted: for example, if the charges of all the particles involved are flipped, or if time is run backwards.

A

'More of the same' might at a glance seem a somewhat dull proposition for a multiverse, but we have to remember that we are talking about *infinitely* more of the same. Within the basic physical parameters, an infinite variety of possible scenarios can play out within different observable Universes. There could be another planet more or less exactly like our own, with another you and another me – except I'm left-handed, you fly to work in an airship and we are both descended from dinosaurs. In fact, if the multiverse really is infinite, there pretty much has to be such a world, alongside every other imaginable and physically possible eventuality.

However, this multiverse of extended spacetime is just *one* type of multiverse – the lowest and most intuitive level among four categorized by influential cosmologist Max Tegmark. In Tegmark's 'mathematical Universe' approach, each type of multiverse can be nested within a higher and more abstruse level.

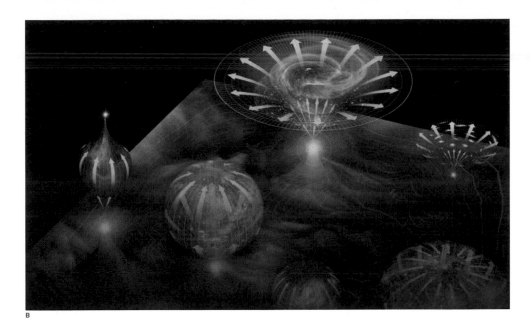

B

Max Tegmark (b. 1967)
Swedish American
Tegmark is a
cosmologist at the
Massachusetts Institute
of Technology, well
known for both his
mathematical Universe
hypothesis and his
work extracting new
information about the
structure of the cosmos
from the CMBR.

So the fairly intuitive 'Level One' multiverse of infinite spacetime just discussed is only one of many such structures (perhaps infinitely many) in a 'Level Two' multiverse, and so on...

But what exactly are these higher levels? They get increasingly abstruse and difficult to grasp as they go higher, so we will start with Level Two.

A   The Level Two multiverse
    idea relies on the possibility
    that new bubbles of
    spacetime with unique
    properties are continuously
    being inflated out of the raw,
    many-dimensional material
    of the cosmos.
B   The precise mix of
    dimensions and physical
    constants in each bubble
    of spacetime determines
    its fate. Some collapse a
    s swiftly as they formed;
    others expand and tear
    apart everything within
    them. Only a small number
    produce the right conditions
    for stable matter.

In essence, a Level Two multiverse is an object that continually gives rise to 'bubbles' that are themselves Level One multiverses, and that may show remarkably *different* characteristics from each other. While all parts of a Level One multiverse share the same essential physics, separate bubbles in a Level Two multiverse could show very different physical constants – and even different arrangements of spacetime with different numbers of dimensions.

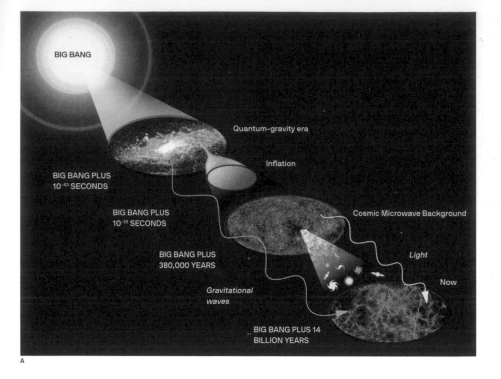

A

The argument for a Level Two multiverse arises from evidence of a crucial event that occurred in the early history of our Universe, only 10^{-35} seconds after the Big Bang itself. Known as Inflation, this event saw a tiny portion of the infant Universe expand suddenly and violently from about the scale of a small atom to the size of a galaxy like the Milky Way, before coming to an end at around 10^{-33} seconds.

The Inflation theory was bolted onto the Big Bang theory by cosmologists including Alan Guth in the United States and Andrei Linde in Russia from the late 1970s onwards, in order to solve a number of discrepancies between the predictions of the original theory and the actual conditions observed in the Universe around us.

Most importantly, it explained the presence of large-scale structure in the Universe, and the very fact that we are here at all. We have already seen that structures had begun to coalesce through the influence of gravity on dark matter even before the formation of the Cosmic Microwave Background Radiation (CMBR), but if the Big Bang conformed to Isaac Newton's cosmological principle (see Chapter 3) it should have produced an absolutely smooth distribution of matter.

So where did the initial slight concentrations of mass, whose enhanced gravity began to attract first dark and then luminous matter into the cosmic web of filaments and voids, originate? Inflation answers this question by suggesting that such seeds resulted from the sudden magnification of tiny variations in the infant Universe inevitable under the rules of subatomic-scale quantum physics. According to the theory, our entire observable Universe, and indeed the wider Level One multiverse beyond it, blew up out of a single atom-sized region within the wider Big Bang. The idea has become widely (if not yet universally) accepted, but obvious questions are: what triggered it, and why did it happen to only our specific part of the Universe?

Alan Guth (b. 1947) Cosmologist Guth is best known for his proposal in 1980 of Inflation as a solution to several issues with the standard Big Bang theory of the time.

Andrei Linde (b. 1948) Russian-American physicist Linde developed ideas about phase transitions in the early Universe that inspired Guth's Inflation theory. Linde subsequently broadened the theory to predict the possibility of an inflationary multiverse.

A Directly after the Big Bang, the Universe experienced a brief era in which the forces were unified in a single superforce that controlled everything. Inflation blew up a small portion of that Universe into everything we know today, but we can only use light to probe the Universe as far back as the CMBR. However, newly discovered 'gravitational waves' could allow astronomers to learn more about conditions in the quantum gravity era.

B The behaviour of subatomic particles is influenced by electromagnetism and the strong and weak forces, but in small numbers they seem immune to gravitation.

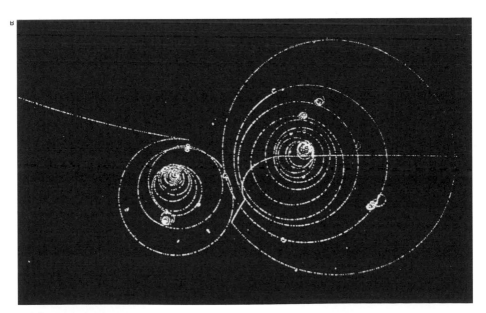

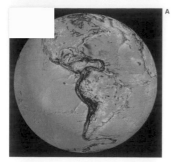

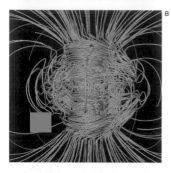

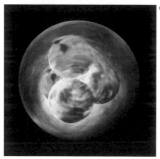

A Gravitation is created by large
   concentrations of mass; this map
   shows the strongest gravity in yellow,
   orange and red.
B Electromagnetic fields are produced
   by particles carrying electric charge.
   Bulk movements of electric current
   in Earth's core produce our planet's
   complex magnetic field. In this shot
   from a computer simulation of the
   generation of the Earth's magnetic
   field (the geodynamo), the rotation
   axis of the modelled Earth is vertical
   and the 3D field is illustrated with
   magnetic lines of force, which are
   coloured blue where the field is
   directed inward and gold where it is
   directed outward. The field is far more
   intense and complicated inside the
   Earth's fluid core where it is generated.
C Both strong and weak forces are far
   stronger than electromagnetism or
   gravity, but limited to incredibly short
   range and therefore only felt inside
   the atomic nucleus.
D Although the forces became
   independent within a fraction
   of a second of the Big Bang,
   physicists can still estimate
   the sequence of splitting.

Cosmologists generally agree that the most plausible driver for Inflation is an event called a 'phase change', which released huge amounts of energy as the Universe's four fundamental forces began to separate out from a primeval 'superforce' in the first instants after the Big Bang. The most familiar phases from our everyday experience are the states of matter – solid, liquid and gaseous arrangements of atoms or molecules such as those seen in ice, water and steam, but for physicists, a phase can be an arrangement of pretty much anything.

Importantly, any phase of a substance has a certain amount of energy bound up in it. Returning to the states of matter, a solid is a low-energy phase in which the individual particles are not moving around much, whereas a vapour is a high-energy phase of fast-moving particles. Transitioning from one phase to another, therefore, involves the supply or release of energy known as latent heat; for example, a pan of boiling water still needs to be supplied with an additional amount of energy (its 'latent heat of vaporization') in order to break the bonds between its molecules and transform into steam, and conversely water cooled to freezing point must get rid of excess energy (its 'latent heat of fusion') in order to form the stable bonds of solid ice.

The most popular explanation for Inflation involves an analogous phase change in the arrangement of the four fundamental forces. A working assumption of modern theoretical physics is that in extreme conditions (high temperatures such as those seen in the first second of the Big Bang, and those briefly generated by collisions in today's particle accelerators), the behaviour governed by these forces starts to become identical. The electromagnetic and weak forces are known to merge into a single 'electroweak' force, and it is thought that the strong force also joins in to create an 'electronuclear' force. Gravitation, as the most peculiar and different of today's four forces, is the most difficult to reconcile, and may only have joined with the others in the first instant of creation itself.

As each force separated out, the initial 'superforce' effectively changed phase and dropped into a less energetic state, releasing a terrific amount of excess energy in the process. It is this energy that is generally assumed to have driven Inflation.

At this stage, you would be forgiven for wondering what exactly Inflation and its driving forces have to do with multi-verses? The answer lies in a radical theory put forward by Linde in 1983: what if Inflation is a natural, ongoing phenomenon, happening everywhere, all of the time?

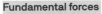
Fundamental forces
The forces that control interactions of matter in the Universe: electromagnetism, gravitation and the strong and weak forces that operate on the tiny scales of the atomic nucleus.

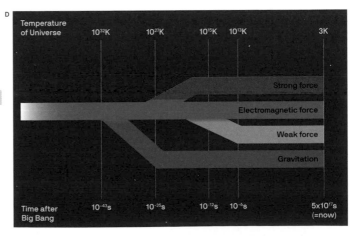

D
Temperature of Universe | 10^{32}K | 10^{27}K | 10^{15}K | 10^{13}K | 3K

Strong force
Electromagnetic force
Weak force
Gravitation

Time after Big Bang | 10^{-43}s | 10^{-35}s | 10^{-12}s | 10^{-6}s | 5×10^{17}s (=now)

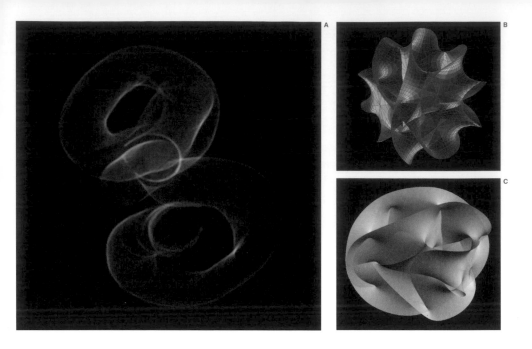

Linde's idea, known as chaotic or eternal inflation, is based on the insight that phases can apply not only to matter and forces, but also to spacetime itself. The same theoretical models that seek to unify the fundamental forces also aim to explain the wide variety of elementary particles that comprise the basis of all matter, and in order to do this they frequently invoke the existence of additional space dimensions.

So-called 'string theories' treat particles as unimaginably small strings of energy vibrating in (usually) ten dimensions, and manifesting different properties depending on the harmonics of their vibration (akin to the different notes produced by a vibrating violin string). Hard as it might be to imagine, the six additional dimensions beyond normal spacetime are extra 'directions' in space, each at right angles to all of the others. We do not perceive these dimensions even at subatomic scales, because they are curled up and wrapped around themselves. From our point of view, just as a ball of string appears as a mere dot when seen from far enough away, so the structure formed by the extra dimensions, known as a 'manifold', appears as a simple point.

The Level Two multiverse is based on the idea that this familiar configuration of dimensions is just one among many possible phases, each with its own characteristic energy level (the same 'vacuum energy' embedded in the fabric of spacetime that is suggested to provide dark energy in our particular Universe). In the right conditions, where a particular phase of spacetime is unstable, new phases can arise spontaneously, rather like bubbles forming in champagne. Depending on the vacuum energy of the new phase relative to its surroundings, the new bubble may be overwhelmed and collapse back upon itself, or it may grow in size. In cases where the vacuum energy of the new phase is large compared to the original phase, the region of new conditions can expand at an ever-accelerating, exponential rate, creating a new inflationary Universe inside the older one.

A   String theories draw a parallel between the vibrational 'modes' of standing waves (e.g. harmonies on a stringed instrument) and the quantized nature of many subatomic properties (e.g. the way particles only have certain values of electric charge).

B   If strings exist, they must be complex objects vibrating in many different dimensions on a scale so small that it can never be observed directly. The extra dimensions are curled and folded in on themselves so as to be imperceptible.

C   This computer visualization attempts to show the way in which a many-dimensional Calabi-Yau manifold (a possible arrangement of higher dimensions in a string) might be perceived by our three-dimensional senses.

D   If string theory is correct, then strings permeate the entire Universe, making up all the matter and energy in the cosmos.

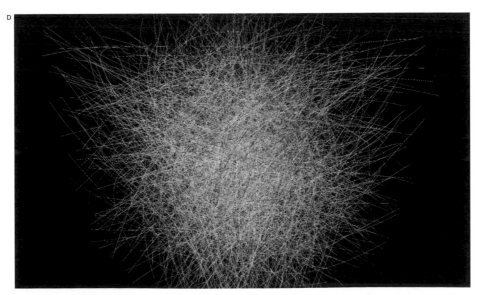

D

A

Linde's idea removes at a stroke the bothersome questions of what came before the Big Bang and how it was triggered. It transforms our Universe (a Level One multiverse) into one of an endless succession, giving rise to each other and sometimes competing for spacetime 'real estate'. While conditions across a Level One multiverse are more or less the same, the separate bubbles in a Level Two structure could display remarkably varied properties, with different numbers of dimensions and wide-ranging values for the fundamental constants of nature.

Of course, the entire theory relies on the as-yet-unproven notion of extra dimensions.

String theories are not the only game in town when it comes to unifying fundamental physics.

A According to M-theory, collisions between neighbouring branes on trillion-year timescales might trigger new Big Bangs.

B As the most distant observable part of the Universe, the CMBR is an obvious place to look for evidence of other Level Two multiverses impinging onto our own.

C A cosmic wake might show up as a ring-like feature in the CMBR, such as that simulated here on the right.

While some of the alternatives also rely on the idea of higher dimensions (for example, the 11-dimensional M-theory), others, such as loop quantum gravity, actively seek to avoid them.

Eternal inflation might sound fantastical, but it is also (at least in theory) provable. If another bubble of spacetime is impinging on our own and is nearby enough to make its impact felt within the observable Universe, it should create visible consequences in the form of a 'cosmic wake' affecting various properties of our Universe, including the distribution of the CMBR. The result, a ring-shaped feature with slightly higher temperature than the average, would be at the limits of current detection technology, but this is likely to change as astronomers continue to probe the relic radiation from the birth of the Universe.

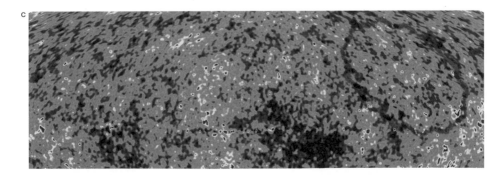

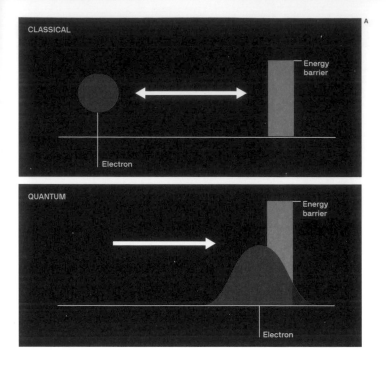

A (Top) In classical physics, the particles emitted by radioactive atomic nuclei should rebound from an insurmountable 'potential barrier' at the edge of the nucleus; radioactive decay should be impossible. (Bottom) The spread-out nature of the quantum wave function creates a small but measurable chance of the particle 'tunnelling' through the potential barrier and emerging on the other side, permitting radioactive decay to occur.

B The quantum wave function is a complex 'object' that exists in at least three dimensions. Like other types of wave, it is subject to interference patterns that can increase the chance of a particle being found in certain places, and lessen the chance of it appearing in others.

Whether or not the Level Two multiverse exists, there is still another yet stranger, higher level of multiverse to come – one that completely unravels any innate concepts of 'shape' that we might still hold onto. The Level Three multiverse arises from the baffling world of quantum physics, and in particular a proposal put forward by Hugh Everett III in the 1950s called the many-worlds interpretation.

Quantum physics is the physics that governs the very small: the realm of subatomic particles. Discovered about a century ago, one of its key insights is wave-particle duality, the idea that objects such as electrons, which we normally think of as particles, can also exhibit wave-like features. This means that precise properties of a particle (such as its position or momentum) can remain wave-like and diffuse until we 'measure' them by observing the particle in certain ways.

One important result of this is to explain the nature of phenomena such as radioactive decay. Individual atoms undergo this transformation unpredictably, but obey rules of probability that describe how many atoms in a particular sample will decay in a certain time period. What is more, the decay process effectively involves particles 'tunnelling' through the energy barrier that holds an atomic nucleus together – something that classical physics says should be impossible, but quantum physics allows by permitting a non-zero chance that the escaping particles appear outside the barrier.

Quantum physics is undeniably real; it is the basis of a great deal of modern technology, including electronics and lasers. But it is also deeply troubling for our understanding of how the Universe works. In our everyday lives, we are used to the predictability of classical physics: things either happen or they don't, with no room for fuzziness of probability except in predicting probable outcomes.

Hugh Everett III (1930–82) US physicist Everett formulated the many worlds interpretation of quantum mechanics for a doctoral dissertation completed in 1956. After his ideas met with fierce resistance, he abandoned academia for industry – it was only in the late 1970s that physicists began to take the many-worlds concept seriously.

Radioactive decay The process in which an unstable nucleus of a particular atom transforms into a more stable state, usually by expelling one or more subatomic particles so that it now forms a different element.

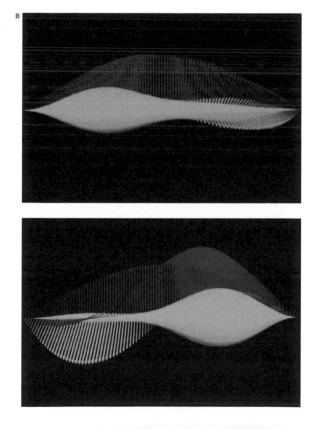

B

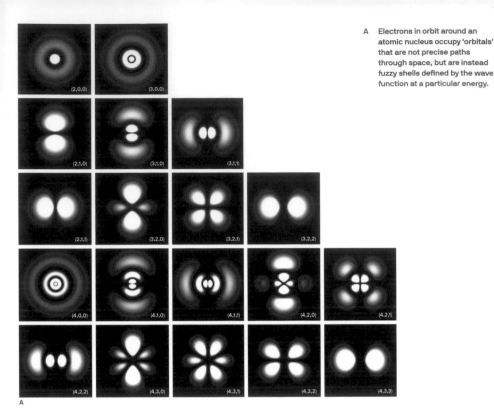

A Electrons in orbit around an atomic nucleus occupy 'orbitals' that are not precise paths through space, but are instead fuzzy shells defined by the wave function at a particular energy.

A

In order to address this point, physicists have come up with a variety of 'interpretations' – rules for bridging the gap between quantum uncertainty and the determinism of the everyday world.

The first and most famous of these is the Copenhagen interpretation, formulated by quantum pioneers Niels Bohr (1885–1962), Werner Heisenberg (1901–76) and others in the mid-1920s. This 'strict' view of quantum uncertainty says that it is a real thing, and the precise properties of particles really are resolved only by observation or measurement.

Schrödinger, developer of the 'wave function' equation at the heart of quantum physics, outlined his doubts about the Copenhagen interpretation in the famous Schrödinger's cat thought experiment. Imagine, he said, that we seal a cat in a box with a small amount of radioactive material, a vial of poison and a mechanism that will release the poison if it detects a radioactive decay event. The experiment is designed so that there is a 50/50 chance of such a decay over its duration, and therefore a 50/50 chance of the cat surviving or being poisoned.

Because the entire experiment hinges on a phenomenon that is subject to quantum rules, Schrödinger argued, if we take the Copenhagen interpretation at face value, the entire system – radioactive sample, poison mechanism and cat – remains in limbo until the moment the box is opened and the radioactive sample is measured (or more broadly, the consequences of its current state are observed). In other words, for the duration of the experiment, the wave function describing the cat is said to exist in a 'quantum superposition', both alive and dead at the same time.

Of course, Schrödinger knew that carrying out such an experiment would be both cruel and pointless, since it is impossible to observe the system 'during' the experiment without bringing it to an end. Nevertheless, he thought that the possibility of a cat trapped between two states was absurd enough to devalue the Copenhagen outlook.

Wave function
An equation describing the wave-like aspect of a quantum particle's properties, such as the probability of its having a particular location or energy

Quantum superposition
A complex wave function created by two or more overlapping wave patterns associated with different possible states of a particle or other quantum system.

B   These pioneers of quantum physics, pictured at the Copenhagen conference of 1930, include (front row left) Niels Bohr and Werner Heisenberg.
C   Schrödinger's well-known thought experiment posits a cat that is both alive and dead at the same time.

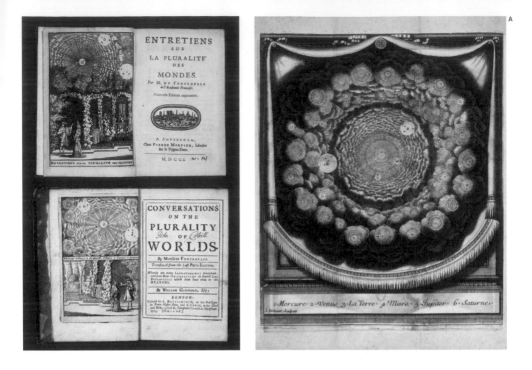

Many physicists have attempted to resolve the problem in different ways, from arguing that the quantum uncertainty somehow naturally resolves itself as its effects ripple out beyond the subatomic scale, to suggesting that the Universe somehow 'knows' the outcome of quantum events in advance, so they are not as uncertain as they appear to be. But Everett's solution is probably the best known, simply because of its mind-boggling implications.

Everett suggested that the different outcomes of quantum events are 'resolved' by the entire Universe splitting onto two divergent paths: one Universe for each possible outcome and its consequences. Schrödinger's cat is never in limbo because the quantum event of radioactive decay itself spins us into one Universe or the other (one where the cat survives and one where it dies). The two Universes tear apart at the speed of light from the location of the event, rather like two leaves of tissue paper gradually being pulled apart.

The many-worlds interpretation, therefore, suggests an infinite number of separate parallel Universes, each of which could be a lower-level multiverse in its own right. Every single quantum-level event since the Big Bang has created its own set of branching Universes. The structure of this Level Three multiverse might, therefore, be compared to a tree, albeit one in which the smallest twigs continue to branch apart an infinite number of times, giving rise to a fractal pattern.

The obvious question, then, is where do the other versions of our Universe exist?

Fractal A mathematical structure such as an equation or geometrical figure that is 'self-similar', with the same patterns recurring again and again on smaller and smaller scales.

A French essayist Bernard de Fontenelle speculated on the possibility of an infinite 'plurality of worlds' in 1686. His observation, 'Behold a universe so immense that I am lost in it... Our world is terrifying in its insignificance' seems particularly fitting when confronted with the unimaginable multiplicity of the multiverse.

B The Mandelbrot set is a well-known fractal: a mathematical description of the boundary created by a deceptively simple equation, which nevertheless shows infinite complexity and 'self-similarity' as we look in more and more detail.

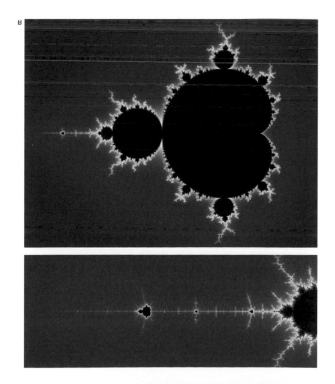

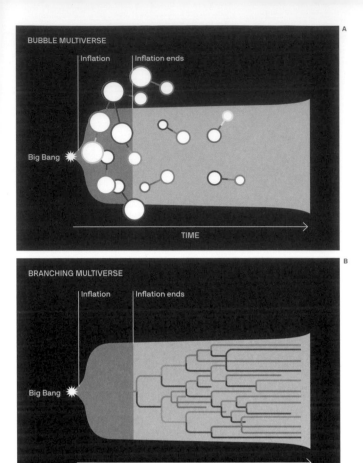

A BUBBLE MULTIVERSE

Inflation Inflation ends

Big Bang

TIME

B BRANCHING MULTIVERSE

Inflation Inflation ends

Big Bang

TIME

A The Level Two multiverse emerges from 'bubbles' of spacetime inflating out of the background material of cosmic foam. While Inflation may end for individual bubble universes, it is still possible for new bubbles to arise from within them if conditions are right.

B The Level Three multiverse is created by continuous branching of different quantum probabilities. Some physicists believe that such a process could not have taken hold until after Inflation had come to an end; if they are right, then such a multiverse, while unimaginably large, would not be truly infinite.

C In 1891, David Hilbert described a remarkable fractal curve that can use a simple set of instructions to fill any finite space with a line of potentially infinite length.

Curiously, the answer is that they exist in exactly the same spacetime that we occupy. In effect, most physicists view the many-worlds interpretation as a statement that our multiverse incorporates all the possible outcomes of quantum events within it; instead of Schrödinger's cat being trapped in a quantum superposition, therefore, it is us who are sitting on one particular branch of an infinitely varied reality (this view can be bolstered by various arguments for why we end up in this particular version of reality – for example, the idea that, since we are here as conscious observers of the Universe, we are necessarily sitting on a branch that has conspired to give rise to our existence).

The wave function describing this sort of multiverse can only be described in terms of a 'Hilbert space', a mathematical structure with a very large (perhaps infinite) number of dimensions. However the dimensions in this case are not usually treated as space dimensions in the sense that we understand them. It is only in the minority 'realist' version of the many-worlds interpretation, which suggests the Universe really does branch to create new physical realities at every point in history, that the multiverse is 'really' a Hilbert space.

The distinction between quantum superpositions and multiple physical realities neatly foreshadows the final and most abstract multiverse of them all: a Level Four multiverse that Tegmark named the 'ultimate ensemble'. This is a mathematical rather than physical object that can give rise to all the other possible types of multiverse.

c

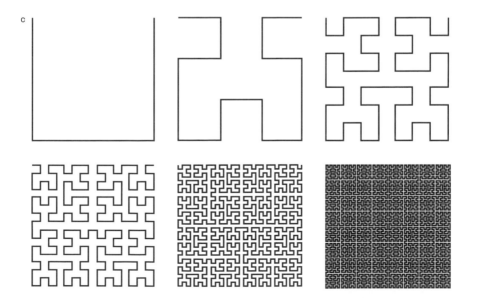

This might seem like a mere box-ticking exercise for mathematicians, but it gives rise to an alarming possibility: if the Universe can be described by a mathematical formula, then could it be calculated by a sufficiently powerful computer? Some scientists and philosophers argue that this could be the case – and furthermore, that if the mathematical simulation was suitably detailed and we were a part of it, we would be unable to distinguish it from the real thing.

A   René Descartes's *Meditations on First Philosophy* (1641) begins by positing the idea of a 'deceiving demon' rendering all our observations of the world around us as illusions. Despite the possibility of such a demon, he shows that the human mind must exist, and goes on to build up a complex metaphysical model of the Universe governed by shifting 'elements'.

B   *The Matrix* (1999) explores the idea of reality as an illusion generated by a sophisticated computer program. It is an intriguing plot for a science-fiction movie, but some philosophers argue that it could be closer to reality than we might care to admit.

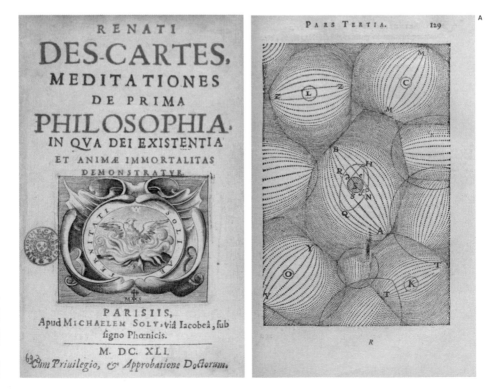

This sort of argument about whether we can really trust our everyday perception of the Universe has been a favourite subject in philosophy for a long time – ever since French philosopher René Descartes (1596–1650) imagined the possibility that a 'deceiving demon' could be providing his senses with the illusion of an apparently external world. More recently, it has been a topic of science-fiction movies such as *The Matrix* (1999).

Shortly after that influential movie's release, philosophers put forward a worryingly convincing argument that we could indeed be in a mathematically simulated multiverse. The idea is simply that if *just one* advanced civilization in the history of the 'real' multiverse develops the ability and interest to run such simulations, then the number of simulated entities in the multiverse will rapidly grow to outnumber the real ones. With no way of telling the difference, we would have to accept that we are probably among the simulations ourselves. Against this, other philosophers and scientists argue that the 'simulation hypothesis' is unscientific since it cannot be disproved, or that there are inherent limits to the capability of computers that would prevent them from ever producing a true simulation of the multiverse.

Whatever the truth, the possibility of multiverses adds vast new levels of complexity to the question of the Universe's shape. It also has important implications for the future development of the cosmos, and our understanding of our own place within it.

Conclusion

Answering the apparently simple question of the shape of space has taken us on a long voyage, from ancient ideas of an Earth-centred Universe to the modern view that we may indeed be the centre of our own *observable* Universe, but that we are otherwise no more significant than a 'grain of sand in the cosmic ocean' as US astronomer Carl Sagan (1934–96) memorably put it.

Along the way, we have seen how gravity plays a key role in shaping spacetime, and have found compelling evidence that the familiar matter of our everyday Universe is vastly outweighed by the unseen mass of dark matter. Yet even the combined gravity of all the mass in the cosmos is not enough to overcome the mysterious dark energy that is pushing space apart at an ever-accelerating rate.

On current evidence, then, today's Universe is 'flat' – essentially uniform in all directions with no large-scale curvature.

What is more, it almost certainly extends far beyond the bounds of our visible Universe into the unreachable realms of an effectively infinite Level One multiverse (regardless of whether the other types of multiverse encountered in Chapter 4 are ultimately proven).

Of course, that may not be the last word on the matter; the past few centuries have seen so many changes to our understanding of the cosmos that few would assume our current state of knowledge is entirely comprehensive. Obvious unanswered questions surround the still-puzzling nature of dark matter and dark energy – in particular, whether the strength of dark energy will continue to grow indefinitely into the future, or whether it might someday peter out or even go into reverse.

A   Do spiral star systems such as our Milky Way galaxy, filled with stars and capable of supporting complex intelligent life, arise by chance, through some guiding influence, or as a result of mechanics of the multiverse?

B   Carl Sagan's books and television series introduced generations to deep questions about the existence of the Universe, life and intelligence.

C   Sagan and fellow scientist Frank Drake designed humanity's first message to the stars, a gold-plated plaque that was carried aboard the Pioneer 10 and 11 spacecraft, launched in the early 1970s and now on their way out of the solar system.

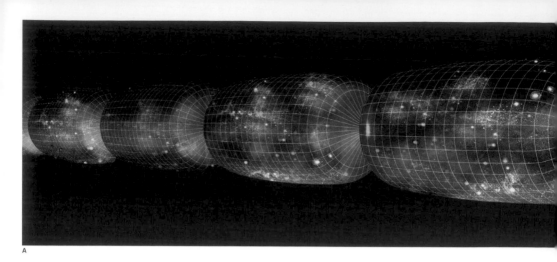

A

There are also rival theories of gravitation that attempt to explain the observed phenomena of the Universe without the need for dark matter or dark energy (although these are very much a minority interest at the moment).

All of this is significant because, while the shape of space might seem like a matter of purely academic interest, it is intrinsically linked to other profoundly important questions concerning both the future development of the Universe and the place of life in general (and ourselves in particular) within it.

In Chapter 3, we explored at length how the density of matter within the Universe, coupled with the strength of dark energy, has the potential to affect both the curvature of spacetime on local levels and the shape of space as a whole. Broadly speaking, this can give rise to a closed spherical Universe, an open saddle-shaped Universe or a flat Universe without any curvature. What has not been mentioned so far, however, is that these shapes determine not only the present geometry of space, but also its likely future evolution.

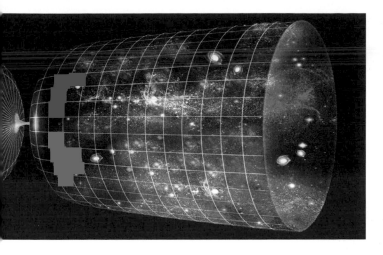

A A cyclic Universe goes through periodic phases of expansion and contraction, with each expansion indistinguishable from the birth of an entirely new Universe.

B The three likely fates of the Universe are that it is open (destined to expand forever, pink), flat (with expansion eventually slowing to a halt, blue), or closed (with expansion eventually reversing, green). Some cosmologists refer to the origin of the Universe in spacetime as the alpha point, and its conclusion (should there be one) as the omega point.

C If the fundamental physical constants of each new Universe can take on truly random values, then the chances of a single Universe fostering the conditions for life are astronomically small.

A closed cosmos is not simply limited today, but is also constrained in its future growth; at some point, gravity will overcome the expansion that started in the Big Bang and begin to pull things back together. In the far future, the Universe will become increasingly dense and its temperature will start to rise, until everything is once again concentrated in a superhot, superdense state known as the 'Big Crunch'. Such a collapse might not happen for trillions of years, and might culminate in spacetime rebounding to create a new Big Bang (potentially a new phase of an unending cyclic Universe), but it would signal a definitive end for the Universe as we know it.

Cyclic Universe
A theory that our Universe is just one in an eternal cycle that will ultimately be replaced by its successor in a new Big Bang.

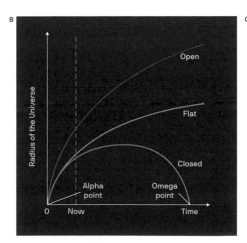

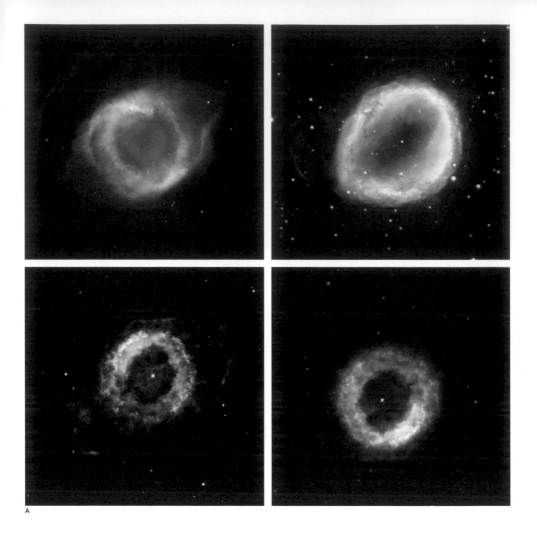

A

Conversely, a present-day open Universe
is one in which gravity will never overcome
the initial speed of expansion – things are
moving apart too rapidly and will never be
drawn back together. Instead, space will
continue to expand forever. In this scenario,
the ultimate fate of the Universe would be
a 'Big Chill'; as galaxies drifted ever further
apart and later generations of stars gradually
exhausted all the viable fuel available for
nuclear fusion, the skies would turn dark,
filled with nothing but slowly cooling
stellar remnants.

A uniform or geometrically flat Universe, superficially similar to the one in which we currently live, was once thought to mark the boundary between these two cosmic fates. Such a Universe was assumed to contain enough mass to slow down expansion, but never quite reverse it. The Universe would continue to grow at an ever-decreasing rate while suffering a similar Big Chill fate to that seen in an open Universe.

However, evidence from measurements of the microwave background suggests that the real reason our Universe is flat is not the mass of matter alone, but the presence of dark energy that contributes to the energy content of space while counter-intuitively driving its expansion. This means that it is possible to have a Universe that is both flat and likely to continue expanding forever without slowing down.

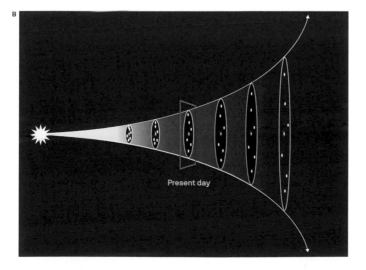

Present day

A   Planetary nebulae are beautiful but short-lived cosmic smoke rings created by dying stars. The burnt-out cores left in their wake continue to glow for billions of years as white dwarf stars, but even these must eventually fade, cool and (ultimately) disintegrate.

B   As dark energy causes space to expand more rapidly in the future, matter and heat energy will be ever more thinly spread, leading the Universe to an inevitable Big Chill.

A

It seems, therefore, that the Universe is almost certainly doomed to a Big Chill, but the discovery that dark energy has apparently increased its strength over cosmic history opens up another most alarming possibility. Instead of slowly cooling and dying over hundreds of billions of years, could dark energy send the cosmos hurtling towards a far earlier date with destiny? The 'Big Rip' hypothesis suggests that the rate of acceleration in cosmic expansion driven by dark energy might not merely be growing steadily, but might instead grow exponentially.

In this scenario, expansion would grow faster and faster over time, overwhelming the attractive force of gravity at first on larger scales and then on smaller and smaller ones. The first signs of this would be the accelerated expansion already detected in the most distant galaxies, but eventually relatively nearby galaxies would also be pulled away by the stretching of space between us and them. Our local Laniakea supercluster would begin to disintegrate, and eventually even our own galaxy would start to come apart as the grip of gravity gave way. Approaching the end, our solar system would fall to pieces, and finally as dark energy became overwhelming, the planets themselves. The final Big Rip would see dark energy overcome the electromagnetic and nuclear forces that bind together molecules and individual atoms, ultimately tearing everything into shreds of subatomic particles.

Exponential growth
A rate of growth that increases in proportion to its current value (hence, for example, the Universe may take x years to double in size, but then only ½x to double again, and ¼x to double again after that).

Laniakea supercluster
The nearest major cosmic structure to the Milky Way, a cloud of galaxies roughly 100 million light years long with our galaxy near one end.

A   Even the densest forms of matter, such as black holes, will evaporate to nothing given sufficient time in a Big Chill cosmos.

B   The Big Rip is an exponential increase in the scale of space due to the growing dominance of dark energy. It first makes itself known on the largest scales, pulling remote galaxies apart at increasing speed (perhaps as we see in the present-day Universe). As it grows in power, it makes itself felt on smaller and smaller scales.

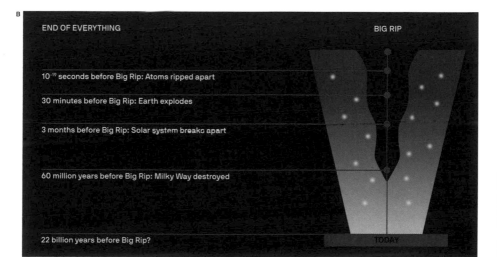

B

END OF EVERYTHING

BIG RIP

10^{-19} seconds before Big Rip: Atoms ripped apart

30 minutes before Big Rip: Earth explodes

3 months before Big Rip: Solar system breaks apart

60 million years before Big Rip: Milky Way destroyed

22 billion years before Big Rip?

TODAY

Constant A number expressing the strength of a relationship between physical properties that is thought to remain the same across the Universe.

Solvent A liquid in which molecules can move around and encounter one another, potentially undergoing chemical reactions.

A   Our planet, Earth, offers seemingly ideal conditions for the evolution of life, but cosmologists wrestle with the question of precisely why this should be so.
B   Life on Earth is so abundant that it has shaped the geological evolution of our planet over billions of years; whole layers of rock strata are composed of the remnants of ancient forests.
C   Experiments such as the Large Hadron Collider have reinforced the message that our cosmos is fine-tuned to allow the existence of stable matter and all that this implies.

Could a Big Rip really happen?

Physicists suggest it is possible only if dark energy takes a specific form called phantom energy that somehow has negative kinetic energy (the 'motion energy' inherent to normal moving objects). In addition to this, the speed of the Rip's growth is dependent on the values of certain cosmological constants, so current measurements suggest that even if such a cataclysm is possible in the first place, it will not happen for many tens of billions of years. Based on our present understanding of the shape and structure of the Universe, our distant descendants are more likely to have to deal with the consequences of a Big Chill than a Big Rip, particularly given that the remaining lifetime of our Sun is just a short few billion years.

Putting aside the question of our ultimate fate, the shape of the Universe also offers a fresh perspective on the biggest question of all – why are we here? One of the most striking observations we can make about the Universe is the fact that it seems suspiciously 'fine-tuned' to foster the development of life.

The properties of the Universe's elementary particles and the strength of the fundamental forces are such that, if things were just slightly different, life might not be possible. Reducing the strength of the electromagnetic force, for example, could dramatically narrow the range of temperatures at which water takes a stable liquid form (something vital for it to act as a solvent for biological chemical reactions). Tamper with gravity and you could prevent planets, stars and galaxies from coalescing at all, or cause them to collapse into black holes. Messing with the nuclear forces, meanwhile, could affect the fusion reactions that power the stars, altering their lifespans or preventing them from shining at all.

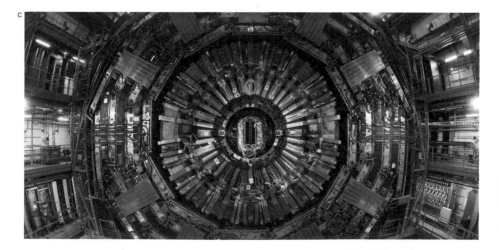

c

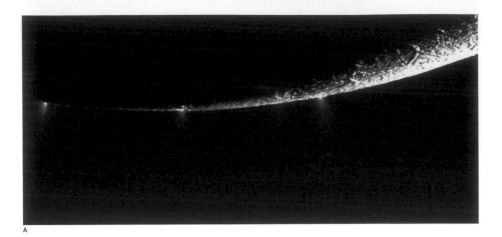

A

Further cosmic coincidences include the dominance of baryonic matter over antimatter, which prevents damaging annihilation reactions from taking place in our region of the Universe (and assures there is matter here *at all*). The quantity of dark matter in the Universe is responsible for the formation of large-scale cosmic structures, and the balance of dark energy, matter and the initial force of cosmic expansion determines the fact that our Universe has evolved in a stable form for billions of years so far, and will continue to do for many billions more.

A Beneath the icy crust of Saturn's small moon Enceladus lie liquid oceans whose water escapes in geyser-like plumes. Such oceans could offer a welcoming environment for alien life in our solar system.

B The discovery of liquid water flowing near the surface of seemingly arid Mars makes this, too, a possible abode of simple life.

In order to explain these apparently amazing coincidences, astronomers turn to a simple but powerful idea called the 'anthropic principle'. Formulated by Brandon Carter in 1973, and later refined by John D. Barrow and Frank Tipler (b. 1947) in the 1980s, the principle comes in 'weak' and 'strong' flavours. The weak anthropic principle is merely the observation that we should not be surprised to find the Universe is fine-tuned to foster the development of life, because if it were not, we would not be here to observe it. The strong anthropic principle (as described by Barrow and Tipler) goes a great deal further, arguing that there is actually an imperative for the Universe to give rise to life – it *really* is fine-tuned in some way.

B

Annihilation The direct conversion of balanced matter and antimatter particles into a burst of energy the moment such particles make contact with each other.

Anthropic Literally 'relating to human beings'. The anthropic principle is essentially the idea that we cannot properly understand the Universe without taking into account our own existence within it.

Brandon Carter (b. 1942) Australian theoretical physicist Carter proposed the first versions of the anthropic cosmological principle in 1973, stating that our own current existence within the Universe needs to be taken into account when assessing our observations of it. He has also done important work on the properties of black holes and neutron stars.

John D. Barrow (b. 1952) British cosmologist Barrow is a professor of mathematical sciences at Cambridge University and a renowned author. Alongside his work on the anthropic principle and the interface between philosophy and physics, he is well known for his statement: 'A universe simple enough to be understood is too simple to produce a mind capable of understanding it.'

Barrow and Tipler put forward three possible causes for a strong anthropic Universe. One is the conscious intervention of an external agency – an extra-dimensional intelligence that created the Universe and set it running in a way that would eventually allow for the development of life. Whether this agent was a deity in the conventional sense, a powerful alien species or a post-human programmer devising a mathematically fine-tuned model (as floated in the simulation hypothesis, see Chapter 4) would barely make a difference except from a theological perspective.

However, the other variants of the strong anthropic idea both rely on fundamental observations about the shape and nature of the cosmos. One is the possibility that the Universe *could not exist* without giving rise to conscious entities capable of observing it. This might sound strange, but think back to the weird world of quantum physics we explored in Chapter 4 – if quantum phenomena really do need to be observed before they resolve themselves into definite outcomes, couldn't the entire Universe similarly be trapped in a state of uncertainty until observed, the cosmic equivalent of Schrödinger's cat?

The idea is all the more intriguing given that this is one possible view of the many-worlds interpretation – the idea that the Universe is held in a vast 'superposition' of countless quantum states that are somehow resolved into the Universe in which we exist. If the conscious observer could be proved to play this vital role, then we might have to accept that we do indeed live in a Level Three quantum multiverse.

A The constellation patterns our ancient ancestors first imposed upon the heavens still symbolize the way in which our view of the Universe is inevitably imposed by our location within it. Plates from *A Celestial Atlas* by Alexander Jamieson, published in 1822.

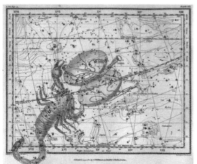

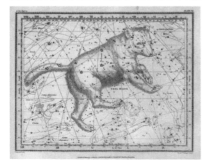

The third variant on the strong anthropic principle, meanwhile, is the idea that our Universe is just one in a vast ensemble that allows every conceivable option to be explored *somewhere*. This is reminiscent of the Level Two multiverse of chaotic inflation – the possibility that the cosmos is an eternal structure giving rise to countless bubble Universes, each with their own unique arrangement of dimensions and other fundamental properties. Many of these bubbles would be vast, empty and lifeless voids, whereas others might be so unstable as to collapse back on themselves as soon as they formed. The fact that 'our' bubble happens to be a habitable one would simply be due to the weak principle.

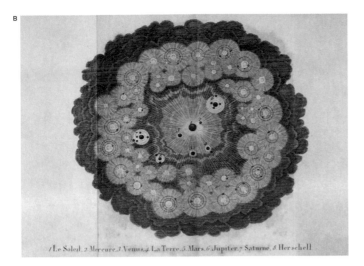

/ Le Soleil, 2 Mercure, 3 Venus, 4 La Terre, 5 Mars, 6 Jupiter, 7 Saturne, 8 Herschell

B   The message of modern
    cosmology is surely
    the same as that which
    Bernard de Fontenelle
    first grasped in the 17th
    century: ours is but one
    among a numberless,
    and perhaps infinite,
    plurality of worlds. Plate
    from *Conversations on
    the Plurality of Worlds* by
    Bernard de Fontenelle,
    published in 1686.

Over the past century, our quest to understand the shape of the Universe has turned out to have implications far beyond anything that Copernicus or even Hubble could have imagined. It has revealed hidden forms of matter and energy, the origin and fate of the cosmos, and perhaps even the secret of life itself. Perhaps the most surprising thing of all is that, given all the exotic possibilities opened up by the secret relationship of gravity and matter, the Universe we can see is about the simplest shape we can possibly imagine.

And yet, if the Universe didn't have that uniform flat structure, we might not be here to see it at all.

Further Reading

Amendola, Luca, *Dark Energy: Theory and Observations* (Cambridge: Cambridge University Press, 2010)

Barrow, John D. and Tipler, Frank J., *The Anthropic Cosmological Principle* (Oxford: Oxford University Press, 1986)

Bartusiak, Marcia, *Einstein's Unfinished Symphony: The Story of a Gamble, Two Black Holes, and a New Age of Astronomy* (New Haven, Connecticut: Yale University Press, 2017)

Davies, Paul, *The Goldilocks Enigma: Why Is the Universe Just Right for Life?* (London: Penguin Books, 2006)

Davies, Paul, *Other Worlds: Space, Superspace and the Quantum Universe* (London: J. M. Dent & Sons, 1980)

Einstein, Albert, *Relativity: The Special and General Theory* (New York: Crown Publishing Group, 1961, English translation)

Gilliland, Ben, *How to Build a Universe: From the Big Bang to the End of the Universe* (London: Philip's, 2015)

Gott, J. Richard and Vanderbei, Robert J., *Sizing up the Universe* (Washington: National Geographic, 2011)

Greene, Brian, *The Elegant Universe: Superstrings, Hidden Dimensions and the Quest for the Ultimate Theory* (London: W. W. Norton and Company, 2003)

Guth, Alan H., *The Inflationary Universe* (New York: Perseus Books, 1997)

Hawking, Stephen, *A Brief History of Time: From Big Bang to Black Holes* (London: Bantam Press, 1988)

Hawking, Stephen, *The Universe in a Nutshell* (London: Transworld Publishers, 2001)

Hoskin, Michael, *The Cambridge Illustrated History of Astronomy* (Cambridge: Cambridge University Press, 1997)

Jones, Mark H., Lambourne, Robert J., and Serjeant, Stephen (eds.), *An Introduction to Galaxies and Cosmology* (2nd revised edition, Cambridge: Cambridge – Open University, 2015)

Krauss, Lawrence M., *The Greatest Story Ever Told – So Far* (New York: Atria Books, 2017)

Krauss, Lawrence M., *A Universe from Nothing: Why There Is Something Rather than Nothing* (New York: Free Press, 2012)

Liddle, Andrew, *An Introduction to Modern Cosmology* (3rd edition, (Chichester, West Sussex: John Wiley and Sons, 2015)

Panek, Richard, *The 4 Percent Universe: Dark Matter, Dark Energy and the Race to Discover the Rest of Reality* (London: Oneworld Publications, 2011)

Rees, Martin, *Just Six Numbers: The Deep Forces that Shape the Universe* (London: Weidenfeld & Nicolson, 1999)

Rovelli, Carlo, *Seven Brief Lessons on Physics* (New York: Riverhead Books, 2016)

Rovelli, Carlo, *Reality Is Not What it Seems: The Journey to Quantum Gravity* (London: Allen Lane, 2016)

Sagan, Carl, *Cosmos* (London: Macdonald and Co, 1980)

Schilling, Govert, *Ripples in Spacetime: Einstein, Gravitational Waves, and the Future of Astronomy* (Cambridge, Massachusetts: Belknap Press, 2017)

Smoot, George and Davidson, Keay, *Wrinkles in Time: Witness to the Birth of the Universe* (New York: William Morrow and Company, 1993)

Sparrow, Giles, *50 Astronomy Ideas You Really Need to Know* (London: Quercus, 2016)

Sparrow, Giles, *The Universe: In 100 Key Discoveries* (London: Quercus, 2012)

Thorne, Kip S., *Black Holes and Time Warps: Einstein's Outrageous Legacy* (London: W. W. Norton and Company, 1994)

Weinberg, Steven, *Cosmology* (Oxford: Oxford University Press, 2008)

Picture Credits

a = above, b = below, c = centre, l = left, r = right

2 DrHitch / Shutterstock.com
4–5 NASA, ESA, H. Teplitz and M. Rafelski (IPAC / Caltech), A. Koekemoer (STScI), R. Windhorst (Arizona State University) and Z. Levay (STScI)
6–7 Andreas Cellarius, *The Harmonia Macrocosmica*, 1660
8–9 ESO / H.H. Heyer
10 Camille Flammarion, *L'Atmosphère: Météorologie Populaire*, 1888
11 Oliver Byrne, *The first six books of the Elements of Euclid*, 1847
12 Babak Tafreshi / National Geographic / Getty Images
13 British Library, Oriental Manuscripts, Add MS 7474 / Qatar Digital Library
14–15 Andreas Cellarius, *The Harmonia Macrocosmica*, 1660
16 Science History Images / Alamy Stock Photo
17 Johannes Kepler, *Mysterium Cosmographicum*, 1596
18–19 T.H. Jarrett (IPAC / SSC)
20 Illustration by Tim Brown
21 NASA, ESA, K. Sahu and J. Anderson (STScI), H. Bond (STScI and Pennsylvania State University), M. Dominik (University of St. Andrews) and Digitized Sky Survey (STScI / AURA / UKSTU / AAO)
21 NASA, ESA, K. Sahu and J. Anderson (STScI), H. Bond (STScI and Pennsylvania State University), M. Dominik (University of St. Andrews)
22 (all) Friedrich Wilhelm Bessel, *Zeichnungen von Halley's Comet 1835*, 1835
23 (both) IndividusObservantis
24 (all) Galileo Galilei, *Sidereus Nuncius*, 1610
25 ESO / S. Brunier
26 William Parsons, *Whirlpool Galaxy*, 1845
27 (all) Cecil G. Dolmage, *Astronomy of To-day*, 1909
28 l Francis Gladheim Pease
28 r NASA, ESA and the Hubble Heritage Team (STScI / AURA)
29 Photo Ardon Bar Hama. © The Albert Einstein Archives, the Hebrew University of Jerusalem, Israel
30 a Case Western Reserve University, Cleveland, Ohio
30 b Illustration by Tim Brown
31 l Hermann Minkowski
31 r, 32 Illustration by Tim Brown
33 a (all) NASA, A. Bolton (UH / IfA) for SLACS and NASA / ESA
33 c NASA, ESA, A. Bolton (Harvard-Smithsonian CfA) and the SLACS Team
33 cb ESA / Hubble and NASA
33 b NASA and ESA
34–35 Jon Morse (University of Colorado), and NASA
36 NASA, ESA, M. Livio (STScI) and the Hubble Heritage Team (STScI / AURA)
37 (all) Isaac Newton, *Opticks: or, A treatise of the Reflections, Refractions, Inflections and Colours of Light*, 1730
38 a Lowell Observatory Archives, Flagstaff, Arizona / Joe Haythornthwaite
38 b Lowell Observatory Archives, Flagstaff, Arizona
39 a Chris North / Cardiff University
39 b Nigel Sharp (NOAO), FTS, NSO, KPNO, AURA, NSF
40 (all) Illustrations by Dan Streat

41 Jon Morse (University of Colorado), and NASA / ESA

42 l NASA / JPL-Caltech

42 r ESA / Hubble and NASA

43 l SSPL / Getty Images

43 r F. W. Dyson, A. S. Eddington, and C. Davidson, *A Determination of the Deflection of Light by the Sun's Gravitational Field, from Observations Made at the Total Eclipse of May 29, 1919*, 1920

44 (all) The SXS (Simulating eXtreme Spacetimes) Project

45 Archives Georges Lemaître, Université catholique de Louvain, Louvain-la-Neuve, Belgium

46 l Special Collections & Archives Department, Nimitz Library, U.S. Naval Academy, Annapolis, Maryland

46 ar, br Illustration by Tim Brown

47 M. Blanton and SDSS

48 l Lars Lindberg Christensen

48 r 2dF Galaxy Redshift Survey

49 Two Micron All-Sky Survey

50 (both) Illustration by Tim Brown

51 l Georges Lemaitre, *The Primeval Atom*, 1950

51 r Fred Hoyle, *The Nature of the Universe*, 1950

52 l FSA / NASA / SOHO

52 r X-ray: NASA / CXC / RIKEN / D.Takei et al; Optical: NASA / STScl; Radio: NRAO / VLA

53 Bettmann / Getty Images

54–55 a NASA / CXC / M. Weiss

55 c COBE Project, DMR, NASA

55 cb NASA / WMAP Science Team

55 b ESA and the Planck Collaboration

56 NASA

57 a Lockheed Missiles and Space Company (Lockheed Martin)

57 bl NASA

57 br NASA / ESA

58 NASA / ESA / Pontificia Universidad Católica de Chile

59 (all) NASA, ESA, A. van der Wel (Max Planck Institute for Astronomy), H. Ferguson and A. Koekemoer (Space Telescope Science Institute), and the CANDELS team

60a NASA / Desiree Stover

60b NASA / MSFC / David Higginbotham

61 a ESA-Planck collaboration

61 b NASA / WMAP Science Team

62 (both) Illustration by Tim Brown

63 Pablo Carlos Budassi

64–65 Science Photo Library/Getty Images

66 al Science Photo Library

66 ac, ar, cl Kevin Rauch / Allan Davis / Science Photo Library

66 c, cr, bl, bc, br Benjamin Bromley / Science Photo Library

67, 68 (all) Illustration by Tim Brown

69 (all) Instituut-Lorentz for Theoretical Physics, Leiden University, The Netherlands

70 National Oceanic and Atmospheric Administration, USA

71 S. Maddox, School of Physics and Astronomy, University of Nottingham et al., APM Survey, Astrophysics, Department of Physics, University of Oxford

72–73 (all) Jason Fletcher, Charles Hayden Planetarium, Museum of Science, Boston

74 NASA / JPL-Caltech / University of Wisconsin

75 a NASA / ESA, The Hubble Heritage Team (STScl / AURA)

75 b NASA, ESA, M. Robberto (Space Telescope Science Institute / ESA) and the Hubble Space Telescope Orion Treasury Project Team

76 al ESA / LFI and HFI Consortia

76 ar MPI

76 bl NASA / ESA / DSS2 (background image); ESA / LFI and HFI Consortia (overlay)

76 br NASA / ESA / DSS2 (background image); MPI (overlay)

77 (both) NASA / JPL-Caltech

78 l NASA, ESA, and the Digitized Sky Survey 2. Davide De Martin (ESA / Hubble)

78 ar NASA, ESA, and the Hubble Heritage Team (STScl / AURA)

78 br Priyamvada Natarajan et al, doi.org/10.1093/ mnras/stw3385

79 a Department of Terrestrial Magnetism, Carnegie Institution of Washington, Washington, D.C.

79 b X-ray: NASA / CXC / MIT / E.-H Peng et al; Optical: NASA / STScl

80 ESO / Digitized Sky Survey 2

Index

References to illustrations
are in **bold**.